Lecture Notes in Mathematics 1551

L. Arkeryd P. L. Lions
P. A. Markovich S. R. S. Varadhan (Eds.)

Nonequilibrium Problems in Many-Particle Systems

Lectures given at the 3rd Session of the Centro Internazionale Matematico Estivo (C.I.M.E.) held in Montecatini, Italy, June 15-27, 1992

Springer-Verlag
Berlin Heidelberg New York
London Paris Tokyo
Hong Kong Barcelona
Budapest

Authors

Leif Arkeryd
Chalmers University of Technology
Department of Mathematics
411296 Göteborg, Sweden

Pierre-Louis Lions
Université de Paris IX, CEREMADE
Place du Général de Lattre de Tassigny
75775 Paris, France

Peter A. Markowich
Technische Universität Berlin
Fachbereich Mathematik
Straße des 17. Juni 136
10623 Berlin, Germany

Srinivasa R. S. Varadhan
NYU-Courant Institute
251 Mercer Street
New York, NY 10012, USA

Editors

Carlo Cercignani
Dipartimento di Matematica
Politecnico di Milano
Piazza Leonardo da Vinci 32
20133 Milano, Italy

Mario Pulvirenti
Dipartimento di Matematica
Università di Roma "La Sapienza"
Piazzale Aldo Moro, 2
00185 Roma, Italy

Mathematics Subject Classification (1991):45K05, 76P05, 82A40, 82A50, 82A70

ISBN 3-540-56945-6 Springer-Verlag Berlin Heidelberg New York
ISBN 0-387-56945-6 Springer-Verlag New York Berlin Heidelberg

Library of Congress Cataloging-in-Publication Data. Nonequilibrium problems in many-particle systems: lectures given at the 3rd Session of the Centro internazionale matematico estivo (C.I.M.E.) held in Montecatini, Italy, June 15-27, 1992 / L. Arkeryd ... [et al.]. p. cm. – (Lecture notes in mathematics; 1551)
ISBN 3-540-56945-6 (Berlin: alk. paper) – ISBN 0-387-56945-6 (New York: alk. paper) 1. Many-body problem–Numerical solutions–Congresses. 2. Nonequilibrium thermodynamics–Congresses. I. Arkeryd, L. (Leif) II. Centro internazionale matematico estivo. III. Series: Lecture notes in mathematics (Springer-Verlag); 1551. QC318.I7N64 1993 530.1'44–dc20 93-5230

Typesetting: Camera-ready by author/editor
46/3140-543210 - Printed on acid-free paper

Preface

This volume contains the text of four sets of lectures delivered by L. Arkeryd, P.-L. Lions, P. A. Markowich and S. R. S. Varadhan at the third session of the Summer School organized by C.I.M.E. (Centro Internazionale Matematico Estivo). These texts are preceded by an introduction written by us which summarizes the present status in the area of Nonequilibrium Problems in Many-Particle Systems (this was the title given to the session) and tries to put the contents of the different sets of lectures in the right perspective, in order to orient the reader.

The lectures presented in this volume deal with the global existence of weak solutions for kinetic models and related topics, the basic concepts of non-standard analysis and their application to gas kinetics, the kinetic equations for semiconductors and the entropy methods in the study of hydrodynamic limits. The lectures were of high level and the school was by all standards a success.

We feel that this volume gives a coherent picture of an important field of applied mathematics which has undergone many important developments in the last few years.

Carlo Cercignani
Mario Pulvirenti

TABLE OF CONTENTS

C. CERCIGNANI - M. PULVIRENTI, Nonequilibrium Problems in Many-Particle Systems. An Introduction 1

L. ARKERYD, Some Examples of NSA in Kinetic Theory 14

P.L. LIONS, Global Solutions of Kinetic Models and Related Problems 58

P.A. MARKOWICH, Kinetic Models for Semiconductors 87

S.R.S. VARADHAN, Entropy Methods in Hydrodynamic Scaling 112

Nonequilibrium Problems in Many-Particle Systems. An Introduction.

C. Cercignani
Dipartimento di Matematica
Politecnico di Milano
Milano (Italy)
and
M. Pulvirenti
Dipartimento di Matematica
Università di Roma "La Sapienza"
Roma (Italy)

1. A sketch of the history of kinetic theory.

According to the atomic theory of matter, all bodies are made up of tiny constituents (particles, molecules, atoms) that, as long as we can ignore quantum effects, move according to the laws of classical mechanics. Thus, *e. g.*, if no external forces, such as gravity, are assumed to act on the particles, each of them will move in a straight line unless it happens to interact with another particle or a solid wall.

Although the rules generating the dynamics of these systems are easy to prescribe, the phenomena associated with this dynamics are not so simple. They are actually rather difficult to understand, especially if one is interested in the asymptotic behavior of the system for long times (ergodic properties) or in the case when the number of particles is very large (kinetic and hydrodynamical limits). Both aspects of the dynamics of molecules are relevant when dealing with a gas, but kinetic theory of gases[1] concentrates upon the problem of outlining the behavior of this system when the number of the particles is very large. This is due to the fact that there are about $2.7 \cdot 10^{19}$ molecules in a cubic centimeter of a gas at atmospheric pressure and a temperature of 0°C.

Given the enormous number of particles to be considered, it would of course be a perfectly hopeless task to attempt to describe the state of the gas by specifying the so-called microscopic state, *i. e.* the position and velocity of every individual particle, and we must have recourse to statistics. This is possible because in practice all that our observations can detect are changes in the macroscopic state of the gas, described by quantities such as density, velocity, temperature, stresses, heat flow, which are related to suitable averages of quantities depending on the microscopic state. At this point, however, a question of principle must be considered. If we knew the exact position and velocity of every molecule of the gas at a certain time instant, the further evolution of the system would be completely determined, according to the laws of mechanics; even if we assume that at a certain moment the positions and velocities of the molecules satisfy certain statistical laws, we are not entitled to expect that at any later time the state of the gas will conform to the same statistical assumptions, unless we prove that this is what mechanics predicts. In certain cases, it turns out that mechanics easily provides the required justification, but things are not so easy, and questions become much more complicated, if the gas is not in equilibrium, as is, *e. g.*, the case for air around a flying vehicle.

Questions of this kind have been asked since the appearance of the kinetic theory of gases; today the matter is relatively well understood and a rigorous kinetic theory is emerging, as the contributions to this volume will illustrate. The importance of this matter stems from the need of a rigorous foundation of such a basic physical theory not only for its own sake, but also as a prototype of a mathematical construct central to the theory of non-equilibrium phenomena in large systems.

As is well known, James Clerk Maxwell (1831-1879) in 1866 developed an accurate method to deal with the nonequilibrium behavior of a gas[2], based on the transfer equations, and discovered the particularly simple properties of a model, according to which the molecules interact with a force inversely proportional to the fifth power of the distance (nowadays commonly called Maxwellian molecules). In the same paper he gave a better justification of a formula that he had previously discovered for the velocity distribution function for a gas in equilibrium.

With his transfer equations, Maxwell had come very close to an evolution equation for the distribution function, but this step must be credited to Ludwig Boltzmann[3] (1844-1906). The equation under consideration is usually called the Boltzmann equation and sometimes the Maxwell-Boltzmann equation (to recognize the important role played by Maxwell in its discovery).

In the same paper, where he gave a heuristic derivation of his equation, Boltzmann deduced an important consequence from it, which later came to be known as the H-theorem. This theorem attempts to explain the irreversibility of natural processes in a gas, by showing how molecular collisions tend to increase entropy. The theory was attacked by several physicists and mathematicians in the 1890's, because it appeared to produce paradoxical results. However, within a few years of Boltzmann's suicide in 1906, the existence of atoms had been definitely established by experiments such as those on Brownian motion.

The paradoxes indicate, however, that some reinterpretation is necessary. Boltzmann himself had proposed that the H-theorem be interpreted statistically; later, Paulus Ehrenfest (1880-1933), together with his wife Tatiana, gave a brilliant analysis of the matter, which elucidated Boltzmann's ideas and made them highly plausible, at least from a heuristic standpoint. A rigorous analysis, however, had still to come.

In the meantime, the Boltzmann equation had become a practical tool for investigating the properties of dilute gases. In 1912 the great mathematician David Hilbert (1862-1943) indicated[4] how to obtain approximate solutions of the Boltzmann equation by a series expansion in a parameter, inversely proportional to the gas density. The paper is also reproduced as Chapter XXII of his treatise entitled *Gründzüge einer allgemeinen Theorie der linearen Integralgleichungen.* The reasons for this are clearly stated in the preface of the book ("Neu hinzugefügt habe ich zum Schluss ein Kapitel über kinetische Gastheorie. [...] erblicke ich in der Gastheorie die glänzendste Anwendung der die Auflösung der Integralgleichungen betreffenden Theoreme").

In about the same year (1916-1917) Sidney Chapman[5] (1888-1970) and David Enskog[6] (1884-1947) independently obtained approximate solutions of the Boltzmann equation, valid for a sufficiently dense gas. The results were identical as far as practical applications were concerned, but the methods differed widely in spirit and detail. Enskog presented a systematic technique generalizing Hilbert's idea, while Chapman simply extended a method previously indicated by Maxwell to obtain transport coefficients. Enskog's method was adopted by S. Chapman and T. G. Cowling in their book

The Mathematical Theory of Non-uniform Gases and thus became to be known as the Chapman-Enskog method.

Then for many years no essential progress in solving the equation came. Rather the ideas of kinetic theory found their way in other fields, such as radiative transfer, the theory of ionized gases and, subsequently, in the the theory of neutron transport. Almost unnoticed, however, the rigorous theory of the Boltzmann equation had started in 1933 with a paper[7] by Tage Gillis Torsten Carleman (1892-1949), who proved a theorem of global existence and uniqueness for a gas of hard spheres in the so-called space homogeneous case. The theorem was proved under the restrictive assumption that the initial data depend upon the molecular velocity only through its magnitude. This restriction is removed in a posthumous book by the same author[8].

In 1949 Harold Grad (1923-1986) wrote a paper[9], which became widely known because it contained a systematic method of solving the Boltzmann equation by expanding the solution into a series of orthogonal polynomials. In the same paper, however, Grad made a more basic contribution to the theory of the Boltzmann equation. In fact, he formulated a conjecture on the validity of the Boltzmann equation. In his words: "From the preceding discussion it is possible to see along what lines a rigorous derivation of the Boltzmann equation should proceed. First, from equilibrium considerations we must let the number density of molecules, N, increase without bound. At the same time we would like the macroscopic properties of the gas to be unchanged. To do this we allow m to approach zero in such a way that $mN = \rho$ is fixed. The Boltzmann equation for elastic spheres, [...] has a factor σ^2/m in the collision term. If σ is made to approach to zero at such a rate that σ^2/m is fixed, then the Boltzmann equation remains unaltered. [...] In the limiting process described here, it seems likely that solutions of Liouville's equation attain many of the significant properties of the Boltzmann equation."

In the 1950's there were some significant results concerning the Boltzmann equation. A few exact solutions were obtained by C. Truesdell[10] in U.S.A. and by V. S. Galkin[11,12] in Soviet Union, while the existence theory was extended by D. Morgenstern[13], who proved a global existence theorem for a gas of Maxwellian molecules in the space homogeneous case. His work was extended by L. Arkeryd[14,15] in 1972.

In the 1960's, under the impact of the problems related to space research, the main interest was in the direction of finding approximate solutions of the Boltzmann equation and developing mathematical results for the perturbation of equilibrium[16,1]. Important methods developed by H. Grad[17] were brought to completion much later by S. Ukai, Y. Shizuta and K. Asano[18-20].

The problem of proving the validity of the Boltzmann equation was still completely open. In 1972, C. Cercignani[21] proved that taking the limit indicated by Grad in the passage quoted above (now currently called the Boltzmann-Grad limit) produced, from a formal point of view, a perfectly consistent theory, *i. e.* the so-called Boltzmann hierarchy. This result clearly indicated that the difficulties of the rigorous derivation of the Boltzmann equation were not of formal nature but were at least of the same order of difficulty as those of proving theorems of existence and uniqueness in the space inhomogeneous case. Subsequently, O. Lanford proved[22] that the formal derivation becomes rigorous if one limits himself to a sufficiently short time interval. The problem of a rigorous, globally valid justification of the Boltzmann equation is still open, except for the case of an expanding rare cloud of gas in a vacuum, for which the difficulties were overcome by R. Illner and M. Pulvirenti[23-24], after that Illner and Shinbrot had provided the corresponding existence and uniqueness theorem for the Boltzmann equation[25].

Recently, R. Di Perna and P. L. Lions[26] have proved a global existence theorem for quite general data, but several important problems, such as proving that energy is conserved or controlling the local growth of density are still open. This result and the ideas related to it will be described in the contribution of P. L. Lions to this volume.

Before this basic result was obtained, the best results in the space inhomogeneous case were those of Arkeryd by means of the techniques of nonstandard analysis. These methods still play an important role in the exploration of open problems of kinetic theory and will be described in the contribution of L. Arkeryd.

The techniques of kinetic theory have become useful in many other fields, such as neutron transport in nuclear reactors, plasma physics, radiative transfer. One of the important recent applications is in the field of semiconductors. As a matter of fact, when the transport of charges in a semiconductor is considered on a sufficiently large time scale, then the motion of the carriers is decidedly influenced by the short range interactions with the crystal lattice, which can be described, in a classical picture of the electron gas, by particle collisions. This situation, which occurs in high-density integrated circuits, explains why there has been an increasing interest in understanding the mathematics of electron transport in submicron structures. The basic tool, in this situation, is given by the Boltzmann equation for carriers, which may exclude the short range interactions between these, which only play a role when the particle density is very large, but can incorporate the Pauli esclusion principle, if necessary. The recent mathematical developments in this field will be reviewed by P. Markowich.

The Boltzmann equation is one of the kinetic equations that can be considered. In dealing with semiconductors, *e. g.*, one frequently considers the field produced by the electrons themselves, which is related to the distribution function through the Maxwell equations because the charge and current density are proportional to moments of the distribution function. This, in the case of no collisions with the lattice, produces the Vlasov-Maxwell system of equations that in the simplest case (quasi steady electric field) reduces to the simpler Vlasov-Poisson system. If collisions are taken into account then one has to face the so called Boltzmann-Vlasov-Maxwell (or Boltzmann-Vlasov-Poisson) system. If the important effect of grazing collisions in a Coulomb field is taken into account, one obtains the so-called Landau equation. Finally, if the quantum aspects of electron transport are also considered one may obtain transport equations from the Schrödinger equation via the Wigner transform. All these equations will appear in the lectures by P. L. Lions and P. Markowich.

So far, we have discussed the Boltzmann (or kinetic) regime and the Vlasov (or mean field) regime for a many particle system in a non-equilibrium situation. It is also of primary importance to illustrate the hydrodynamical behavior of such a system. Real fluids are usually described by the Euler (or Navier-Stokes) equations, which are believed to be a reduced description of the particle system. Actually, if the right space-time scales are adopted and an appropriate limit is taken, Newton equations formally lead to the Euler equations. Unfortunately, very little is known on this problem from a rigorous point of view. Some progress, however, has been recently achieved by the so called entropy method as will be illustrated in the lectures of R. S. Varadhan. Most of his analysis will be devoted to model systems which are ruled by a stochastic dynamics.

2. Basic equations and properties.

The Boltzmann equation is an evolution equation for the distribution function $f(x, \xi, t)$, which gives the probability density of finding a molecule at position x at time t with velocity ξ. If we assume that there is no body force (such as gravity) acting on the particles, we may write the Boltzmann equation in the following form

$$\frac{\partial f}{\partial t} + \xi \cdot \frac{\partial f}{\partial x} = Q(f, f) \tag{2.1}$$

where

$$Q(f,f) = \int \int (f' f'_* - f f_*) B(\xi - \xi_*, n) d\xi_* dn \tag{2.2}$$

Here $B(\xi - \xi_*, n)$ is a kernel associated with the details of the molecular interaction, f' f'_*, f_* are the same thing as f, except for the fact that the argument ξ is replaced by ξ', ξ'_*, ξ_*. The latter is an integration variable having the meaning of the velocity of a molecule colliding with the molecule of velocity ξ, whose evolution we are following, while ξ' and ξ'_* are the velocities of two molecules entering a collision which will bring them into a pair of molecules with velocities ξ and ξ_*. n is a unit vector defining the direction of approach of two colliding molecules. For details, we refer to the bibliography[1]. The collision term, although complicated, has many interesting properties, such as

$$\int Q(f,f)\phi(\xi) d\xi = \frac{1}{4} \int f f_* (\phi' + \phi'_* - \phi - \phi_*) B(|\xi - \xi_*|, n) d\xi_* d\xi dn \tag{2.3}$$

We now observe that the integral in Eq. (2.3) is zero independent of the particular functions f and g, if

$$\phi + \phi_* = \phi' + \phi'_* \tag{2.4}$$

is valid almost everywhere in velocity space. Since the integral appearing in the left hand side of Eq. (2.2) is the rate of change of the average value of the function ϕ due to collisions, the functions satisfying Eq. (2.4) are called "collision invariants". They play an important role in the discussion of the Boltzmann equation. It can be shown[1,27,28] that the most general solution of Eq. (2.4) is given by

$$\phi(\xi) = A + B \cdot \xi + C|\xi|^2 \tag{2.5}$$

Another important result is obtained by letting $\phi = \log f$ in Eq. (2.3). In fact the properties of the logaritmic function lead to the ***Boltzmann inequality:***

$$\int_{R^3} \log f Q(f,f) d\xi \leq 0 \tag{2.6}$$

Further, the equality sign applies if, and only if, $\log f$ is a collision invariant, or, equivalently:

$$f = A\exp(-\beta|\xi - v|^2) \tag{2.7}$$

where A is a positive constant related to a, c, $|b|^2$ (β , v, A constitute a new set of constants). The function appearing in Eq. (2.7) is the so called *Maxwell distribution* or *Maxwellian*. It is a simple corollary, then, that the Maxwellians are the only functions for which $Q(f, f)$ vanishes.

3. The Vlasov equation and the mean-field limit.

In this section we exploit a few elementary facts concerning the Vlasov dynamics and its relation to the Newton equations.

Our starting point is a conservation law of the type:

$$\partial_t f(x,t) + \operatorname{div}(uf(x,t)) = 0 \tag{3.1}$$

where $f = f(x,t)$ is a probability density, $x \in \mathbf{R}^N$ and t is time; u is a vector field:

$$u = u(x,t) \in \mathbf{R}^N \tag{3.2}$$

which is a linear functional of f of the form

$$u(x,t) = \int dy K(x-y) f(y,t) \tag{3.3}$$

where K is a given, smooth, vector-valued kernel.

Consider now the N-particle system obeying the following ordinary differential system:

$$\frac{dx_i}{dt} = \frac{1}{N} \sum_{j=1}^{N} K(x_i - x_j) \tag{3.4}$$

and the measure-valued function

$$\mu^N(t, dx) = \frac{1}{N} \sum_{j=1}^{N} \delta(x - x_j(t)) \tag{3.5}$$

where $x_j(t)$, $j = 1, 2, \ldots, N$ is a solution of Eq. (3.4) and δ denotes, as usual, the Dirac measure. If we let

$$\mu^N(t, \phi) = \frac{1}{N} \sum_{j=1}^{N} \phi(x_j(t)) = \int \mu^N(t, dx)\phi(x) \tag{3.6}$$

where ϕ is a bounded smooth function, an easy calculation shows that

$$\frac{d}{dt}\mu^N(t, \phi) = \mu^N(t, u \cdot \nabla\phi) \tag{3.7}$$

In other words $\mu^N(t, \cdot)$ is a weak solution to Eq. (3.1).

The following natural question arises. Assume that, at time zero

$$\mu^N(0,\phi) \to \int f(0,x)\phi(x)dx, \text{ as } N \to \infty \quad ; \tag{3.8}$$

then is the same true at time t? *I. e.*, denoting by $f(x,t)$ the solution of Eq. (3.1) with initial datum $f(0,x)$, we ask whether

$$\mu^N(t,\phi) \to \int f(t,x)\phi(x)dx, \text{ as } N \to \infty \quad . \tag{3.9}$$

This convergence can indeed be proved. Actually, it is nothing else than a continuity property of the solutions of Eq. (3.1) with respect to initial conditions, in the topology of the weak convergence of measures.

Thanks to (3.9) we can say that Eq. (3.1) has been rigorously derived, in the so called mean-field approximation, starting from the particle dynamics (3.4).

The above analysis can be slightly modified to include the Vlasov equation, which, in conservation form, reads as follows:

$$\partial_t f(x,\xi,t) + \operatorname{div}_{x,\xi}(Uf(x,\xi,t)) = 0 \tag{3.10}$$

where:

$$U = (\xi, K * \rho) \in \mathbf{R}^{2d} \qquad (\rho = \int f(x,\xi)d\xi) \tag{3.11}$$

where d (=2,3) is the number of dimensions of the physical space and * denotes convolution.

So far, we have assumed that K is smooth. An important case for applications is, however, the kernel

$$K(x) = \alpha \frac{x}{|x|^{d-1}} \tag{3.12}$$

where $\alpha \in \mathbf{R}$ is a constant. In this case Eq. (3.10) is called Vlasov-Poisson for obvious reasons. It describes a gas of charged particles (or plasma) in the mean field approximation.

Magnetic effects can also be considered. In this case we are led to the Vlasov-Maxwell equations which can be studied in a relativistic framework as well.

Due to the singularity of the kernel (3.12), the validity of the Vlasov-Poisson equation has not been established as yet and the mere existence and uniqueness of smooth solutions in dimension 3 has only recently been achieved.

An analysis of the existence theory of the Vlasov-Poisson and Vlasov-Maxwell equations will be presented by P. L. Lions, while a practical application of these equations in the context of the semiconductor theory will be illustrated by P. Markowich.

4. Kinetic theory and fluid dynamics.

In this section we compare the microscopic description supplied by kinetic theory with the macroscopic description supplied by continuum gas dynamics. To this end we introduce the definitions of density ρ, of bulk velocity v (with components v_i), of random velocity c, of stress tensor with components p_{ij}, of heat flow vector q:

$$\rho = \int_{R^3} f d\xi \tag{4.1}$$

$$v = \frac{\int_{R^3} \xi f d\xi}{\int_{R^3} f d\xi} \tag{4.2}$$

$$c = \xi - v \tag{4.3}$$

$$p_{ij} = \int_{R^3} c_i c_j f d\xi; \qquad (i,j = 1,2,3) \tag{4.4}$$

$$\rho e = \frac{1}{2} \int_{R^3} |c|^2 f d\xi; \tag{4.5}$$

$$q_i = \frac{1}{2} \int_{R^3} c_i |c|^2 f d\xi \tag{4.6}$$

Then, using the fact that 1, ξ_i and ξ^2 are collision invariants, we multiply the Boltzmann equation by these functions and integrate with respect to ξ, to obtain:

$$\frac{\partial \rho}{\partial t} + \sum_{i=1}^{3} \frac{\partial}{\partial x_i}(\rho v_i) = 0, \tag{4.7}$$

$$\frac{\partial}{\partial t}(\rho v_j) + \sum_{i=1}^{3} \frac{\partial}{\partial x_i}(\rho v_i v_j + p_{ij}) = 0, \qquad (j = 1,2,3) \tag{4.8}$$

$$\frac{\partial}{\partial t}(\frac{1}{2}\rho|v|^2 + \rho e) + \sum_{i=1}^{3} \frac{\partial}{\partial x_i}[\rho v_i(\frac{1}{2}|v|^2 + e) + \sum_{j=1} v_j p_{ij} + q_i] = 0. \tag{4.9}$$

These equations are the balance equations of mass, momentum and energy well-known in continuum mechanics. It is not worthless to mention, at this point, that Eqs. (4.7-9) are *not* fluid-dynamical equations. Actually they cannot even be solved without first solving the Boltzmann equation to determine p_{ij} and q_i. There are situations. however, where the distribution function can be shown to be very close to a Maxwellian, so that q_i and the anisotropic part of p_{ij} are negligible, and, by taking

$$q_i = 0, \qquad p_{ij} = p\delta_{ij}, \tag{4.10}$$

we can describe the gas by means of the Euler equations. How to pass from the kinetic regime (described by the Boltzmann equation) to the hydrodynamical regime (described by the Euler equations) is one of the interesting problems related with the Boltzmann equation, that we shall now sketch.

5. Scaling properties.

A point of great relevance in the study of the Boltzmann equation is the analysis of the scaling properties: a large system, as we shall presently see, can be more conveniently described in terms of fluid-dynamical equations, when it is considered on a suitable space-time scale.

Let us consider a gas obeying the Boltzmann equation, confined in a large box Λ_ϵ of side ϵ^{-1}, ϵ being a parameter to be sent to zero. Let $f^\epsilon = f^\epsilon(x,\xi,t), x \in \Lambda_\epsilon$ be the number density of the particles. We assume that the total number of particles is proportional to the volume of the box, *i. e.* we normalize f^ϵ as follows:

$$\int_{\Lambda^\epsilon \times \mathbf{R}^3} f^\epsilon(x,\xi)dxd\xi = \epsilon^{-3} \tag{5.1}$$

We also assume that the time evolution is given by the Boltzmann equation

$$\frac{\partial f^\epsilon}{\partial t} + \xi \cdot \frac{\partial f^\epsilon}{\partial x} = AQ(f^\epsilon, f^\epsilon) \tag{5.2}$$

and look at the behavior of the system on the scale of the box; in this case we have to use appropriate space and time variables, because in terms of the variable x, the box is of size ϵ^{-1}, while we would like to regard it as of order unity. Thus we introduce the new independent and dependent variables

$$r = \epsilon x, \qquad \tau = \epsilon t; \qquad (r \in \Lambda) \tag{5.3}$$

$$\hat{f}(r,\xi,t) = f^\epsilon(x,\xi,t) \tag{5.4}$$

Clearly, $\hat{f}$ describes the gas on the scale of the box and is normalized to unity:

$$\int_{\Lambda \times \mathbf{R}^3} \hat{f}(r,\xi)drd\xi = 1 \tag{5.5}$$

The picture of the (same) system in terms of the variables r and τ is called *macroscopic*, while the picture in terms of x and t is called *microscopic*. Note that on the macroscopic scale the typical length for the kinetic phenomena described by the Boltzmann equation, *i. e.* the mean free path, turns out to be of order ϵ (since it is of order unity on the scale described by x). Thus sending the size of the box to infinity like ϵ^{-1} or the mean free path to zero like ϵ are equivalent limiting processes.

In terms of the macroscopic variables, Eq. (5.2) reads as follows:

$$\frac{\partial \hat{f}}{\partial \tau} + \xi \cdot \frac{\partial \hat{f}}{\partial r} = \epsilon^{-1} AQ(\hat{f},\hat{f}) \tag{5.6}$$

Thus, on the scale of the box, the mean free path (inversely proportional to the factor in front of Q) is reduced by a factor ϵ. This means that the average number of collisions diverges when $\epsilon \to 0$ and the collisions become dominant. For Eq. (5.6) to hold,

$Q(\hat{f}, \hat{f})$ must be small of order ϵ, so that $\hat{f}$ is expected to be close to a Maxwellian, whose parameters are, in general, space and time dependent. In this case the macroscopic balance equations (5.7- 9) can be closed through Eqs. (5.10) to obtain the Euler equations for a perfect compressible fluid. These considerations can be made rigorous and appropriate references will be given below.

For the present time, let us mention other physical considerations concerning our scaling. To this end, let us consider a small portion of fluid in a neighborhood of a point $r \in \Lambda$ (fig. 1). By the scaling transformation this portion is magnified into a large system of particles, which is seen to evolve on a long time scale. It will have a tendency to "thermalize" so that its distribution will quickly become a local Maxwellian with parameters $A(\epsilon^{-1}r)$, $\beta(\epsilon^{-1}r)$, $v(\epsilon^{-1}r)$ suitably related to the fluid-dynamical fields ρ , e, v. These will evolve according to the Euler equation on a much slower scale of times.

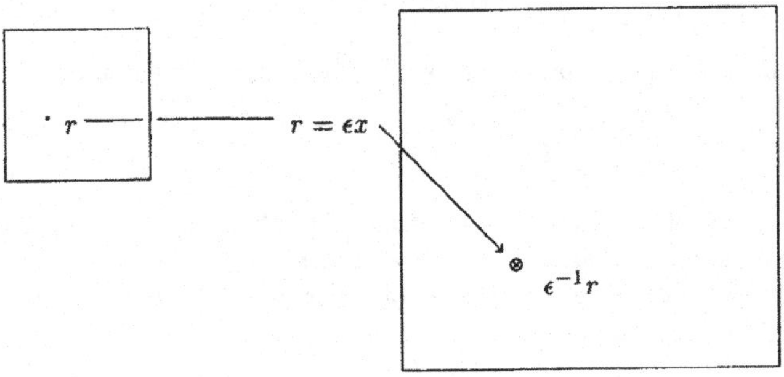

Fig. 1

Thus we have illustrated two different time scales. The fast one, which we call *kinetic*, is of the order of the time necessary to reach a local equilibrium, a process described by the Boltzmann equation. The slow scale, which we call *fluid-dynamical*, describes the time evolution of the parameters of the local Maxwellian.

It would be even more natural to apply the same considerations to the Newton equations. But, although one might expect that such dynamics should yield, under the above scaling, the Euler equations, our ignorance of the long time behavior of the Hamiltonian systems is such that, at the moment, we are quite far from a rigorous derivation of the equations of hydrodinamics from the basic laws of Classical Mechanics. As we shall see in the lectures by R. S. Varadhan, however, the hydrodynamics of a class of Hamiltonian systems can be derived if we assume that some ergodic properties are satisfied, at least as far as a smooth solution of the Euler equations exists.

It may be worth, at this point, to underline how different is the hydrodynamic behavior of a gas obeying the Boltzmann equation and thus the state law of perfect gases, from the behavior arising from a particle system describing a real gas and thus a more complicated state law, including the effects of the interaction potential between molecules. In other words, as a consequence of the Boltzmann-Grad limit, the local equilibrium of a Boltzmann gas is that of a free gas, while, in general, the local equilibrium of a gas is a Gibbs state for an interacting particle system. Although the latter is the local equilibrium taking place in real fluids, the mathematical analysis of the hydrodynamics arising from the Boltzmann equation is technically easier and has produced more results.

Let us now analyse another scaling, which clarifies the nature of the Boltzmann-Grad limit. We now require the number of particles in Λ_ϵ to be of the order of ϵ^{-2}, *i. e.* we replace Eq. (5.1) by

$$\int_{\Lambda^\epsilon \times \mathbf{R}^3} f^\epsilon(x,\xi)dxd\xi = \epsilon^{-2} \tag{5.7}$$

In order to keep the normalization to unity of $\hat{f}(r,\xi,t)$, expressed by Eq. (5.5) we change the scaling from Eq. (5.4)

$$\hat{f}(r,\xi,t) = \epsilon^{-1} f^\epsilon(x,\xi,t) \tag{5.8}$$

Then we obtain, in place of Eq. (5.6)

$$\frac{\partial \hat{f}}{\partial \tau} + \xi \cdot \frac{\partial \hat{f}}{\partial r} = \mathcal{A}Q(\hat{f},\hat{f}) \tag{5.9}$$

Hence the Boltzmann equation is invariant for the space-time scaling (5.3), provided that the particle number goes as the power 2/3 of the volume. This invariance property suggests that the Boltzmann equation can be derived from the BBGKY hierarchy via a space time scaling with the total number of particles proportional to ϵ^{-2}; this is what can be checked at a formal level[1,21]. It is also clear why the Boltzmann-Grad limit is frequently called the ***low density*** limit; in fact, in this limit, the particle number in a large box divided by the volume of the box goes to zero. The number of collisions per unit (macroscopic) time stays finite, while it diverges in the hydrodynamical limit, as we saw before.

We summarize the content of this discussion in the graph below.

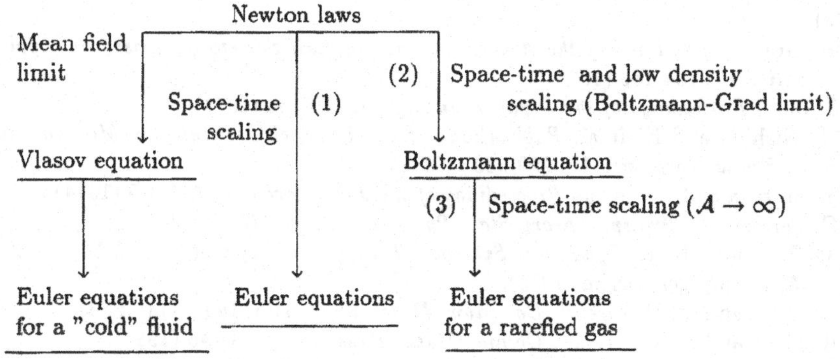

As we said before the limit corresponding to arrow (1) is not completely understood as yet and the best attempts in this direction will be discussed in the lectures by R. S. Varadhan. The limit corresponding to arrow (2) has been proved for short times and globally only for an expanding rare cloud of gas. The limit corresponding to arrow (3) is well understood for times up to the occurence of the first singularity in the fluid-dynamical equations[29,30,31,32]. The mean field limit (4) is well understood for smooth bounded interaction potentials[33,34,35].

Finally, it is also possible to derive (again before shocks develop) a hydrodynamical regime for the Vlasov dynamics (arrow (5))[36,37]. We also mention that the limit (3) is just one of a large class for which the incompressible Euler and Navier-Stokes equations can also be derived[38,39,40].

REFERENCES

1. C. Cercignani, *The Boltzmann equation and its applications*, Springer, New York (1988).

2. J. C. Maxwell, *Philosophical Transactions of the Royal Society of London* **157**, 49-88 (1867).

3. L. Boltzmann, *Sitzungsberichte Akad. Wiss.*, Vienna, part II, **66**, 275-370 (1872).

4. D. Hilbert, *Mathematische Annalen* 72, 562-577 (1912).

5. S. Chapman, *Proceedings of the Royal Society* (London) **A93**, 1-20 (1916/17).

6. D. Enskog, *Kinetische Theorie der Vorgänge in mässig verdünnten Gasen, I. Allgemeiner Teil*, Almqvist & Wiksell, Uppsala (1917).

7. T. Carleman, *Acta Mathematica* **60**, 91-146 (1933).

8. T. Carleman, *Problèmes Mathématiques dans* la *Théorie Cinétique des Gaz*, Almqvist & Wiksell, Uppsala (1957).

9. H. Grad, *Comm. Pure Appl. Math.* **2**, 331-407 (1949).

10. C. Truesdell, *Jour. Rat. Mech. Anal.* **5**, 55-128 (1956).

11. V. S. Galkin *PMM (in Russian)* **20**, 445-446 (1955).

12. V. S. Galkin, *PMM* **22**, 532-536 (1958).

13. D. Morgenstern, *Proceedings of the National Academy of Sciences* (U.S.A.) **40**, 719-721 (1954).

14. L. Arkeryd, *Arch. Rat. Mech. Anal.* **45**, 1-16 (1972).

15. L. Arkeryd, *Arch. Rat. Mech. Anal.* **45**, 17-34 (1972).

16. C. Cercignani, *Mathematical Methods in Kinetic Theory*, Plenum Press, New York (1969).

17. H. Grad, *Proceedings of the American Mathematical Society Symposia on Applied Mathematics* **17**, 154-183 (1965).

18. S. Ukai, *Proceedings of the Japan Academy* **50**, 179-184 (1974).

19. T. Nishida and K. Imai, *Publications of the Research Institute for Mathematical Sciences, Kyoto University* **12**, 229-239 (1977).

20. Y. Shizuta and K. Asano, *Proceedings of the Japan Academy* **53**, 3-5 (1974).

21. C. Cercignani, *Transp. Theory Stat. Phys.*, 211-225 (1972).

22. O. Lanford, III, in *Dynamical Systems, Theory and Applications*, J. Moser, Ed., **LNP 35**, 1, Springer, Berlin (1975).

23. R. Illner and M. Pulvirenti, *Commun. Math. Phys.* **105**, 189-203 (1986) .

24. R. Illner and M. Pulvirenti, *Comm. Math. Phys.* **121**, 143-146 (1989).

25. R. Illner and M. Shinbrot, *Comm. Math. Phys.*, **95**, 217-226 (1984).

26. R. Di Perna and P. L. Lions, *Ann. of Math.* **130**, 321-366 (1989).
27. C. Cercignani, *J. Stat. Phys.* **58**, 817-824 (1990).
28. L. Arkeryd and C. Cercignani, *Rend. Mat. Acc. Lincei s. 9*, **1**, 139-149 (1990).
29. R. Caflisch, *Comm. Pure Appl. Math.* **33**, 651-666 (1980).
30. M. Lachowicz, *Math. Meth. in Appl. Sci.* **9**, 342-366 (1987).
31. T. Nishida, *Comm. Math. Phys.* **112**, 119-148 (1978).
32. S. Ukai and K. Asano, *Hokkaido Math. J.* **12**, 303-324 (1983)
33. W. Braun, K. Hepp, *Comm. Math. Phys.* **56**, 101-120 (1977).
34. R. L. Dobrushin, *Sov. J. Funct. Anal.* **13**, 115-119 (1979)
35. H. Neunzert, in *Kinetic Theories and the Boltzmann Equation*, C. Cercignani, ed., 60-110, **LNP 1048**, 207, Springer, Berlin (1984).
36. K. Oelschläger, *Arch. Rat. Mech. Anal.* (to appear, 1993).
37. S. Caprino, R. Esposito, R. Marra and M. Pulvirenti, *Comm. PDE* (to appear, 1993).
38. A. De Masi, R. Esposito and J. L. Lebowitz, *Comm. Pure Appl. Math.* **42**, 1189-1214 (1989).
39. C. Bardos, F. Golse, D. Levermore, *J. Stat. Phys.* **63**, 323-344 (1991).
40. C. Bardos, F. Golse, D. Levermore, *Fluid dynamical limits of kinetic equations II. Convergence proof for the Boltzmann equation.* Preprint.

Some examples of NSA methods in kinetic theory

L. Arkeryd

As you all know, the interesting behaviour in nonlinear evolution situations is usually too violent to be controlled mathematically by the much used tool of contracting mappings. Routinely various analytic, algebraic, geometric and topological ideas are then brought in to tackle the problem in question. In these lectures I will try to draw your attention to a *different line* of approach, namely Robinson's nonstandard analysis or NSA. As an added bonus, NSA is usually suitable for merging with the first mentioned classical types of ideas. Moreover, from the point of view of physics, what happens at distances or within volumes below, say, the scale of elementary particle phenomena, is an artefact of the model with little experimental relevance. So, also in that perspective, the question whether our evolution model starts from an underlying set of rationally, really, or infinitesimally spaced points, should be decided purely on mathematical grounds.

So why NSA? You are all familiar with the observation that it was infinitesimal methods which brought about the rich mathematical harvest of the 18th century. And in present day mathematics our reasoning can often be considerably simplified using NSA-arguments -- continuity properties turned into finite counting arguments, advanced stochasticity reasoning into simple combinatorics, and convergence in measure arguments into a mere distinguishing between infinitesimal, finite and infinite magnitudes. There is also the deeper, general fact, as stressed by J. Keisler [K], that 'almost all of classical mathematics uses only a small part of ZFC, at most second order arithmetic with Π_1^1 comprehension, whereas Robinson's analysis uses principles beyond Π_1^1 comprehension in a natural way, thus bringing more of ZFC within the reach of our intuition'.

On a series of problems from gas kinetics, I will now try to illustrate some advantages in tackling nonlinear evolution equations using NSA. My five examples -- from the Carleman model and the Broadwell model, via time dependent and stationary problems for the full Boltzmann equation, to a validation question -- rely in an increasingly strong way on methods from NSA. They are chosen to illustrate the way of making a first penetration of a standard problem via NSA, the use of NSA as a technical tool in proving standard mathematical results, and the use of NSA to create a suitable statistical physics setting for a problem. An easy access to the examples will assume knowledge of NSA on the level of an introductory text into the subject, such as [AFHL], [HL] or [L]. Each example ends with a comment on the nonstandard approach.

References

[AFHL] Albeverio, S., Fenstad, J.E., Høegh-Krohn, R., Lindstrøm, T., *Nonstandard methods in stochastic analysis and mathematical physics*, New York, Acad. Press, 1986.

[HL] Hurd, A., Loeb, P., *An introduction to nonstandard real analysis*, New York, Acad. Press, 1985.

[K] Keisler, J., The hyperreal line, *Synthese*, to appear.

[L] Lindstrøm, T., An invitation to nonstandard analysis, in *Nonstandard analysis and its applications*, ed. N. Cutland, U.K., Cambridge Univ. Press, 1988.

Example 1 (The Carleman equation, C.E.)

As our first application of NSA to kinetic theory, we shall consider the simplest velocity discrete model, the C.E. on the full real line. It describes the evolution of two (non-negative) functions u and v depending on one space variable $x \in R$ and on time $t \in R_+$,

$$(\partial_t + \partial_x)u = v^2 - u^2,$$

$$(\partial_t - \partial_x)v = u^2 - v^2.$$

We shall discuss this equation with non-negative initial values,

$$u(x,0) = u_0(x), \quad v(x,0) = v_0(x)$$

for $x \in \mathbb{R}$. In particular, an existence theorem will be proved for the case of the initial data being measures. This obviously poses a problem, since the right hand side is not defined, when u and v are only measures.

We start with a quick sketch of the classical theory for the C.E.. Evidently given $x \in R$ and $t > 0$ a solution at (x,t) depends only on the previous values in the triangular domain between (x,t), $(x-t,0)$ and $(x+t,0)$, as can be seen from the integrated form of the C. E.. The right hand side of the equation is loc. lip. in C_0^1, so

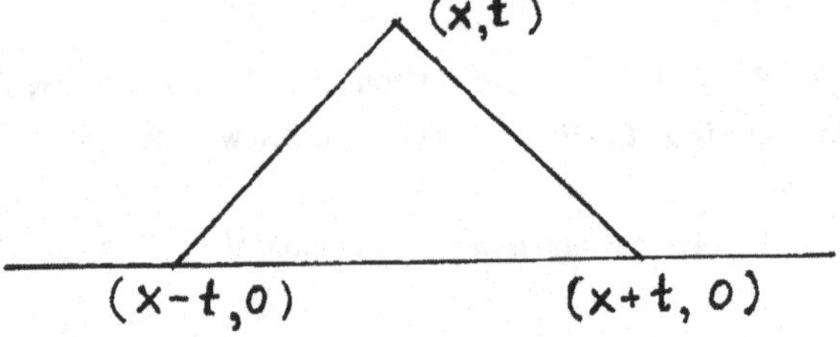

(e.g. using contracting mapping arguments on the mild form) there is a unique local (in time) solution for given initial values in C_0^1. It is non-negative (since an iterative construction in the exponential form gives non-negative solutions) and conserves mass (by integration with respect to x of the sum of the two equations.)

The contracting mapping argument also works for initial values in $L^1 \cap L^\infty$, and the solution will then depend continuously in L^1 on the initial values if they are uniformly bounded in L^∞. There is also a comparison lemma, which goes back to Kolodner.

Lemma 1.1. [Ko] If $U_0 \geq u_0 \geq 0$, $V_0 \geq v_0 \geq 0$ are initial values (in C_0^1) with (unique) solutions (in C_0^1) for $0 \leq t \leq \epsilon$, then on this time interval

$$U_t \geq u_t, \quad V_t \geq v_t.$$

Proof Take $(u_1, v_1) = (U-u, V-v)$. Then

$$(\partial_t + \partial_x)u_1 + (U+u)u_1 = (V+v)v_1,$$

$$(\partial_t - \partial_x)v_1 + (V+v)v_1 = (U+u)u_1,$$

$$u_1(0) = U_0 - u_0 \geq 0, \quad v_1(0) = V_0 - v_0 \geq 0.$$

By a contracting mapping argument this problem has a unique solution. It is non-negative by a construction similar to the exponential iteration construction already mentioned.

The reasoning also works for $U = V = \max_x(u_0 \vee v_0)$.

Corollary 1.2. [Ko] $\max_x (u(x,t) \vee v(x,t)) \leq \max_x (u_0(x) \vee v_0(x))$.

The corollary implies that the previous local solution can be continued to a global one uniformly bounded by $\|u_0\|_\infty \vee \|v_0\|_\infty$.

Kolodner's maximum bound was improved by Illner and Reed (in C_0^1).

Lemma 1.3. [IR] $\max (u(x,t) \vee v(x,t)) \leq C/t$, $t > 0$, with C only depending on the total mass $\int_{-\infty}^{\infty} (u_0(x) + v_0(x))dx$.

For the (slightly technical) proof see [IR].

By transfer the above results for the C.E. also hold in the NSA context. In particular, if φ is a standard C_0^∞-mollifier (i.e. $\varphi \geq 0$, $\int \varphi = 1$), then with $\varphi_\epsilon(x) = \frac{1}{\epsilon}\varphi(\frac{x}{\epsilon})$, $\epsilon \approx 0$, $u_0, v_0 \in L_0^1$ (standard), it holds that

$$\int |\varphi_\epsilon * u_0 - u_0| + |\varphi_\epsilon * v_0 - v_0| \, {}^*dx \approx 0,$$

and so (by continuity) $(u_\epsilon(\cdot,t), v_\epsilon(\cdot,t)) \approx (u(\cdot,t), v(\cdot,t))$ in L^1-sense. (We employ the usual convention of writing u_0, v_0 and φ for $*u_0$, $*v_0$ and $*\varphi$ also in the nonstandard context.) Moreover,

$$\varphi_\epsilon * u_0, \ \varphi_\epsilon * v_0 \in {}^*C_0^1,$$

and so

$$u_\epsilon(\cdot,t), \ v_\epsilon(\cdot,t) \leq C/t, \quad t > 0$$

with C only depending on $\int(u_0 + v_0)dx$. It follows that

$$u_t, v_t \leq C/t, \ t > 0,$$

in a.e. sense. Moreover, if u_0, v_0 are positive measures with compact support, then

$$u_{\epsilon 0}, v_{\epsilon 0} \in {}^*C_0^1,$$

and still there is a NSA solution $u_\epsilon(\cdot,t)$, $v_\epsilon(\cdot,t)$ in ${}^*C_0^1$ bounded by C/t. In particular, this solution is finite for ${}^o t > 0$, and $({}^o u_\epsilon(\cdot,t), {}^o v_\epsilon(\cdot,t))$ is a mild Loeb $L_+^1 \cap L_+^\infty$ solution in the sense that for Loeb a.e. $x \in ns^*R$

$$^o u_\epsilon(x+t_2, t_2) = {}^o u_\epsilon(x+t_1,t_1) + \int_{t_1}^{t_2} ({}^o v_\epsilon^2 - {}^o u_\epsilon^2)(x+t,t)Ldt \tag{1.1}$$

for $0 < {}^o t_1 < {}^o t_2 < \infty$, and for $\psi \in C_0(\mathbb{R})$

$$\lim_{t\to 0} \int {}^o u_\epsilon(x,t)^{o*}\psi(x)Ldx = \int \psi(x)du_0(x),$$

together with the analogous statements for ${}^o v_\epsilon$.

Notice that our Loeb solutions are $\mathbb{R}$-valued defined for $x \in ns^*R$. We shall prove that they correspond to Young measure solutions (which are probability measure valued defined for $x \in R$). For this we insert a brief discussion of Young measure solutions, cf [C].

Let $\operatorname{supp} u_{\epsilon t} \cup \operatorname{supp} v_{\epsilon t} \subseteq [-\lambda,\lambda]$ for $t \le t_2$. For $0 < t_1 < t_2$ and D a Borel set in $[-\lambda,\lambda] \times [t_1,t_2] \times [0, \frac{C}{t_1}]$, introduce the measure

$$Q_{t_1,t_2}(D) = L\{y \in {}^*[-\lambda,\lambda] \times {}^*[t_1,t_2];\ ({}^o y, {}^o u_\epsilon(y)) \in D\}.$$

Let $B \subseteq [0, \frac{C}{t_1}]$ be a Borel set, $\mathscr{A} \subseteq [-\lambda,\lambda] \times [t_1,t_2]$ be a Lebesgue measurable set.

Then

$$Q_{t_1,t_2}(\mathscr{A} \times [0, \tfrac{C}{t_1}]) = \text{Lebesgue measure of } \mathscr{A},$$

and so for B fixed, $Q_{t_1,t_2}(\cdot \times B)$ is Lebesgue absolutely continuous. Hence there is a measurable function $U_{.}(B)\colon y \to U_y(B)$, such that

$$Q_{t_1,t_2}(\mathcal{A} \times B) = \int_{\mathcal{A}} U_y(B)dy.$$

Moreover,

1) $U_y([0, \frac{C}{t_1}]) = 1,$

2) $U_y(\bigcup_j(B_j) = \sum_j U_y(B_j),$ if the B_j's are disjoint,

3) $U_y(\phi) = 0.$

This implies that

i) U_y is a probability measure on the Borel sets of $[0, \frac{C}{t_1}]$,

ii) $Q_{t_1,t_2}(\mathcal{A} \times B) = \int_{\mathcal{A}} U_y(B)dy.$

Such a mapping $y \to U_y$ is a Young measure. Moreover, the above can be considered also for fixed t, $t_1 \leq t \leq t_2$, $y = (x,t)$, giving $U_x = U_{xt} = U_y$. Finally, for $g(\zeta) = \zeta^2 \in \mathbb{R}$, set $g(U_y) = \int_0^{C/t_1} g(\zeta)dU_y(\zeta)$. The same constructions can be made starting from v_ϵ giving a Young measure $y \to V_y$.

Lemma 1.4. $\int {}^{0*}\phi(y, {}^0u_\epsilon(y))Ldy = \int \phi(y, U_y)dy$, if $\phi \in C([-\lambda,\lambda] \times [t_1,t_2] \times [0, \frac{C}{t_1}])$, and analogously for v_ϵ and V.

Proof When y is near y_0, then

$$\sup_{0 \leq \zeta \leq C/t_1} |\phi(y,\zeta) - \phi(y_0,\zeta)|$$

is small, so it is enough to establish the lemma for $\phi(y,\zeta) = \psi(\zeta)$ when $y \in K$ and $\phi = 0$ when $y \notin K$. Here K is an arbitrary axis-parallel square in

$[-\lambda,\lambda] \times [t_1,t_2]$.

Since ψ can be uniformly approximated by stepwise constant functions, it is enough to prove the lemma when $\psi(\zeta) = \psi_j$ = constant on $[\zeta_j, \zeta_{j+1})$, and $\bigcup_j [\zeta_j, \zeta_{j+1}) = [0, C/t_1]$. Now with

$$\mathcal{A}_j = \{y \in {}^*K;\ {}^o u_\epsilon(y) \in [\zeta_j, \zeta_{j+1})\}$$

we have

$$\int_K \psi(U_y)dy = \sum_j \psi_j \int_K U_y([\zeta_j, \zeta_{j+1}))dy = \sum_j \psi_j \int_{\mathcal{A}_j} Ldy =$$

$$= \int_{{}^*K} \psi({}^o u_\epsilon(y))Ldy.$$

Using Lemma 1.4 it follows that (U_y, V_y) defined from ${}^0(u_\epsilon,v_\epsilon)$, the Loeb solution (1.1), satisfies the weak Young measure equation

$$\int_{-\lambda}^{\lambda} \int_0^{C/t_1} dU_y(\zeta)\phi\Big|_{t=t_2} dx =$$

$$= \int_{-\lambda}^{\lambda} \int_0^{C/t_1} dU_y(\zeta)\phi\Big|_{t=t_1} + \int_{t_1}^{t_2} dt \int dx \int_0^{C/t_1} D_t\phi(x+t,t)dU_{(x+t,t)}(\zeta) + \tag{1.2}$$

$$+ \int_{t_1}^{t_2} \int_{-\lambda}^{\lambda} \phi(y)(g(V_y) - g(U_y))dtdx, \quad \phi \in C^1([-\lambda,\lambda]\times[t_1,t_2],$$

and the analogous equation with V_y and U_y interchanged.

In particular, if (u,v) is a standard $L^1 \cap L^\infty$ solution, then $U_y = \delta_{u(y)}$, $V_y = \delta_{v(y)}$, and the Young measure solutions (equation) is a distribution solution (equation). One interest in the Young measure solutions lies in their describing the limiting behaviour of standard approximations, say $(u_{\epsilon_j}, v_{\epsilon_j})$ with $0 < \epsilon_j \in \mathbb{R}^+$, when $\epsilon_j \searrow 0$.

Theorem 1.5. There is a Loeb solution $({}^o u_\epsilon, {}^o v_\epsilon)$ of type (1.1) for $\epsilon \approx 0$, with corresponding Young measure $(U_\cdot, V_\cdot)$ satisfying (1.2), if and only if there is a standard sequence $(\epsilon_j)_{j \in \mathbb{N}}$, $\epsilon_j \searrow 0$, such that the corresponding weak C.E.'s converge to this Young measure equation.

Proof The if-part: By a saturation argument there is an $\epsilon = \epsilon_{j_0} \approx 0$ ($j_0 \in {}^*N_\infty$), such that $(u_{\epsilon_{j_0}}, v_{\epsilon_{j_0}})$ on each standard test function is infinitesimally close to the limit. The corresponding Young measures $(U_\cdot, V_\cdot)$ satisfy by Lemma 1.4.

$$\int \phi(y, U_y)dy = \int^{o*} \phi(y, {}^o u_\epsilon(y))Ldy = {}^o\int^* \phi(y, u_\epsilon(y))^*dy =$$

$$= {}^o\int {}^*\phi(y,\zeta)dyd\delta_{u_\epsilon(y)}(\zeta) = \lim_{j\to\infty} \int \phi(y,\zeta)dyd\delta_{jy}(\zeta),$$

where

$$\delta_{jy} = \delta_{u_{\epsilon_j}(y)} \text{ and } \phi \in C([-\lambda,\lambda] \times [t_1,t_2] \times [0, C/t_1]),$$

and analogously for $V_\cdot$.

For the 'only if'-part another lemma is needed. Let $\mu(\cdot)$ be the Lebesgue measure on $[-\lambda,\lambda] \times [t_1,t_2]$ and $\delta_{u_\epsilon(\cdot)}$ the Dirac measure at $u_\epsilon(\cdot)$ on $\zeta \in [0, \frac{C}{t_1}]$. Let W be the closure of the set of measures $\{\mu(\cdot)\delta_{u_\epsilon(\cdot)};\ 0 < \epsilon < 1\}$ defined on $\mathscr{A} = [-\lambda,\lambda] \times [t_1,t_2] \times [0, C/t_1]$, with respect to the weak* topology (generated by $C(\mathscr{A})$-functions), and let d be the corresponding metric. Then W is compact. We need a well-known approximation of ${}^*W$ depending on W being compact, and for convenience also recall the proof.

Lemma 1.6. If $\tilde{\mathcal{E}} \in {}^*W$, then there is $\mathcal{E} \in W$, such that ${}^*d(\mathcal{E}, \tilde{\mathcal{E}}) \approx 0$, i.e. $\tilde{\mathcal{E}}$ is in the monad of $\mathcal{E}$.

Proof Suppose the contrary, i.e. given $a \in W$, there is $\eta_a > 0$ in $\mathbb{R}^+$ such that $\tilde{\mathcal{E}} \notin \{F \in {}^*W;\ {}^*d(a,F) < \eta_a\}$. But $W \subseteq \bigcup_{a \in W} \{F;\ d(a,F) < \eta_a\}$, so by compactness there is a finite subcovering

$$W \subseteq \bigcup_{\ell=1}^{\nu} \{F;\ d(a_\ell,F) < \eta_{a_\ell}\},$$

and so

$${}^*W \subseteq \bigcup_{1}^{\nu} \{F;\ {}^*d(a_\ell,F) < \eta_{a_\ell}\},$$

which gives the contradiction.

Proof Only if: Given u_ϵ, $0 < \epsilon \approx 0$, there is by Lemma 1.6 an $\mathcal{E} \in W$, such that

$${}^*d(\mathcal{E}, \mu(\cdot)\delta_{u_\epsilon(\cdot)}) \approx 0.$$

And so there is a standard sequence $(\epsilon_j)_{j \in \mathbb{N}}$, $\epsilon_j \searrow 0$, such that in weak* measure sense $\mu(\cdot)\delta_{u_{\epsilon_j}(\cdot)} \rightharpoonup \mathcal{E}$, and for $\phi \in C([-\lambda,\lambda] \times [t_1,t_2] \times [0, \frac{C}{t_1}])$

$$\lim_{j \to \infty} \int \phi(y,\zeta) dy d\delta_{jy}(\zeta) = \int \phi(y,\zeta) d\mathcal{E}(y,\zeta) =$$

$$= {}^{\circ}\!\int^* \phi(y,\zeta) {}^*d\mu(y) d\delta_{u_\epsilon(y)}(\zeta) =$$

$$= \int {}^{\circ *}\phi(y, u_\epsilon(y)) L d\mu(y) = \int \phi(y,U_y) dy.$$

Here the last step followed by Lemma 1.4.

Remark i) If $u_0 = u_{0\ell} + \sum_1^\infty a_k \delta_{x_k}$, $v_0 = v_{0\ell} + \sum_1^\infty a_k \delta_{x_k}$, and $u_{0\ell}$, $v_{0\ell}$ are Lebesgue integrable, then there is a solution in $L^1 \cap L^\infty$ for $t > 0$, which in the simplest case $u_0 = v_0 = \delta_0$ is given by

$$u(x,t) = \frac{t+x}{t^2} \chi_{(-t,t)}(x), \quad v(x,t) = u(-x,t), \quad t > 0.$$

This is Wick's solution [W]. It is an open problem, if there are solutions in $L^1 \cap L^\infty$ for $t > 0$, also for measure-valued initial conditions in general.

ii) Consider the case of a velocity discrete Boltzmann equation (for simplicity on a torus Λ) with microreversibility, physical consistency, and conservation of momentum and energy. If the initial values are Loeb integrable with finite entropy, then there is a (global in time) mild Loeb L^1 solution to the initial value problem. The solution conserves mass and has a globally bounded H-function. Alternatively, if the initial values are standard L^1 with finite entropy, there are Young measure solution. Also, there is a theorem similar to Theorem 1.5 giving a 1-1 correspondence between Loeb L^1 solutions and Young measure solutions. There is not yet a correspondingly general global L^1 existence theory for the discrete Boltzmann equation. The proofs are similar to the above ones for the C.E.

iii) It seemed natural for me to start considering the C.E., with the initial data being measures, in the above non-standard framework of hyperreal-valued functions. That made it possible to use suitable classical results directly, and delivered the Loeb solutions more or less automatically. Recalling previous Loeb L^1 studies of the Boltzmann equation, the Young measure interpretation was immediate. Of course, once it is known that Young measure solutions with decay estimates are relevant for the C.E. when the initial data are measures, then one may just as well study them directly within a standard framework. So in this example, one may argue that the main advantage with NSA, is to facilitate the initial analysis of the problem.

References

[C] Cutland, N., Internal controls and relaxed controls, J. London Math. Soc. 27, 130-140 (1983).

[IR] Illner, R. and Reed, M.C., The decay of solutions to the Carleman model, *Math. Meth. Appl. Sci. 3*, 121-127 (1981).

[Ko] Kolodner, I.I., On the Carleman model, *Ann. Mat. pura appl.* 63, Series 4, 11-32 (1963).

[W] Wick, J., Some remarks on the Carleman model, Tagungsbericht 47/1982, Oberwolfach, also in *Math. Meth. Appl. Sci.* 6, 515 (1984).

Example 2 (The Broadwell model)

This example is based on joint work with R. Illner [AI], and considers the long time behaviour of solutions to the Broadwell model in one space dimension

$$\begin{aligned} (\partial_t + \partial_x)v &= z^2 - vw \\ (\partial_t - \partial_x)w &= z^2 - vw \\ \partial_t z &= \tfrac{1}{2}(vw - z^2), \end{aligned} \tag{2.1}$$

$v = v(t,x)$, $w = w(t,x)$, $z = z(t,x)$, $t \in [0,\infty)$, $x \in [0,1]$, with initial condition

$$\begin{aligned} v(0,x) &= v_0(x) \geq 0 \\ w(0,x) &= w_0(x) \geq 0 \\ z(0,x) &= z_0(x) \geq 0, \end{aligned} \tag{2.2}$$

and boundary condition

$$\begin{aligned} v(t,0) &= w(t,0) \\ v(t,1) &= w(t,1), \end{aligned} \tag{2.3}$$

for $t > 0$. We shall confine our attention to nonnegative bounded continuous initial values, but the results generalize to $v_0, w_0, z_0 \in L_+^\infty$. The following results are well known.

Theorem 2.1. If v_0, w_0, and z_0 are nonnegative and continuous, then (2.1-3) has a global nonnegative continuous solution.

For a proof see e.g. [B].

Theorem 2.2. The global solution given by Theorem 2.1 satisfies

$$\frac{d}{dt}\int_0^1 (v(t,x) + w(t,x) + 4z(t,x))dx = 0, \tag{2.4}$$

(mass conservation)

$$\frac{d}{dt}\int_0^1 (v-w)(t,x)dx = 2v(t,0) - 2v(t,1), \tag{2.5}$$

(momentum transfer)

$$\frac{d}{dt}\int_0^1 (v+2z)(t,x)dx = v(t,0) - v(t,1),$$

$$\frac{d}{dt}\int_0^1 (w+2z)(t,x)dx = -w(t,0)+w(t,1), \tag{2.6}$$

and

$$\int_0^1 (v\log v + w\log w + 4z\log z)(t,x)dx + \int_0^t\int_0^1 (vw-z^2)\log\frac{vw}{z^2}(\tau,x)dxd\tau$$

$$= \int_0^1 (v_0\log v_0 + w_0\log w_0 + 4z_0\log z_0)dx \tag{2.7}$$

(H-theorem).

It is easy to obtain the unique steady solutions of (2.1), which are expected in the limit $t \to \infty$. By time independence from $\partial_t z = 0$, we get $vw = z^2$, and so $\partial_x v = \partial_x w = 0$. The boundary condition (2.3) implies that v, w and z must all equal the same constant $a > 0$, and from the mass conservation law (2.4), $a = \int_0^1 (v_0 + w_0 + 4z_0)dx/6$. An obvious guess is now that the above time dependent solutions converge to a.

Theorem 2.3. [AI] Let $g(t,x)$ denote any of the three functions $v(t,x)$, $w(t,x)$ or $z(t,x)$. Then

$$\lim_{t\to\infty}\int_0^1 |g(t,x) - a|dx = 0. \tag{2.8}$$

In the standard context

$$\int_0^\infty dt \int_0^1 (z^2 - vw)\log \frac{z^2}{vw}\, dx < \infty,$$

which in the NSA context implies

$$\int_{t_1}^{t_2} {}^*dt \int_0^1 (z^2 - vw)\log \frac{z^2}{vw}\, {}^*dx \approx 0 \quad (t_1, t_2 \in {}^*R_\infty^+). \tag{2.9}$$

This, however, immediately leads to the following idea for a proof. From (2.4) and (2.9) it follows that $z^2 - vw \approx 0$ outside of some infinitesimal subset of $\mathcal{M} = {}^*[0,1] \times [t_1, t_2]$. Inserting this into (2.1) and integrating along the characteristics one formally obtains that -- modulo infinitesimals -- v, w and z are constant along the characteristics. Now the geometry gives (see figure, dotted lines are characteristics)

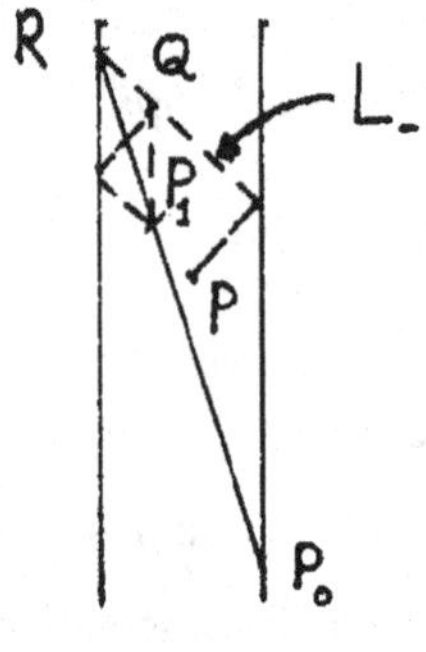

$z^2(Q) \approx v(Q)w(Q)$, $z^2(P_1) \approx v(P_1)w(P_1)$, $z(Q) \approx z(P_1)$, $v(Q) \approx w(P_1)$, $v(P) \approx w(Q)$,

and so

$$v(Q)v(P) \approx v(Q)\, v(P_1) \tag{2.10}$$

hence

$$w(Q) \approx v(P) \approx v(P_1). \tag{2.11}$$

We can thus conclude that modulo infinitesimals, v, w and z are equal to a in $\mathcal{M}$ outside of an infinitesimal set, and so

$$\int |v_t - a| + |w_t - a| + |z_t - a|^* dx \approx 0, \quad t \in {}^*R^+_\infty.$$

This is equivalent to (2.8).

For a strict proof along these lines, the following two lemmas will be used. Since an integration along a characteristic of z^2-vw may pick up noninfinitesimal mass from an infinitesimal set where $z^2-vw \not\approx 0$, we will work with renormalized solutions. This requires the following consequence of (2.9).

Lemma 2.4. Let $t_1 \in {}^*R^+_\infty$, $t_2 = t_1 + 5$ in the above definition of $\mathcal{M}$, and $g = v, w$ or z. Then

$$\int_{\mathcal{M}} \left| \frac{z^2 - vw}{1 + g} \right|^* dxdt \approx 0.$$

Proof We prove the lemma for $g = w$. The other cases are similar. Set

$$\mathcal{M}_1 = \{(x,t) \in \mathcal{M}; z^2 \leq 2\, vw\}.$$

By (2.9)

$$\int_{\mathcal{M}\setminus\mathcal{M}_1} \frac{|z^2-vw|}{1 + w}{}^* dxdt \leq \frac{1}{\log 2} \int_{\mathcal{M}} (z^2 - vw)\log \frac{z^2}{vw}{}^* dxdt \approx 0.$$

To estimate the corresponding integral over $\mathcal{M}_1$, we set

$$\mathcal{M}_2 = \{(x,t) \in \mathcal{M}_1; vw > n^{-1}, |\tfrac{z^2}{vw} - 1| \geq n^{-1}\},$$

$$\mathcal{M}_3 = \{(x,t) \in \mathcal{M}_1; vw \leq n^{-1}\},$$

$$\mathcal{M}_4 = \{(x,t) \in \mathcal{M}_1; |\tfrac{z^2}{vw} - 1| < n^{-1}\}.$$

Evidently,

$$\int_{\mathcal{M}_2} \frac{|z^2 - vw|}{1+w} {}^*dxdt \le \int_{\mathcal{M}_2} v^*dxdt.$$

Now on $\mathcal{M}_2$, $|z^2 - vw| \ge \frac{vw}{n}$, and $|\log \frac{z^2}{vw}| \ge \frac{1}{2n}$, and so

$$\frac{1}{2n^3} \int_{\mathcal{M}_2} {}^*dxdt \le \int_{\mathcal{M}} (z^2 - vw)\ln \frac{z^2}{vw} {}^*dxdt \approx 0.$$

So for small enough $n \in {}^*\mathbb{N}_\infty$,

$$\int_{\mathcal{M}_2} {}^*dxdt \approx 0,$$

hence by the entropy estimate (2.7)

$$\int_{\mathcal{M}_2} \frac{|z^2 - vw|}{1+w} \le \int_{\mathcal{M}_2} v^*dxdt \approx 0.$$

On $\mathcal{M}_3$, $z^2 \le 2vw \le \frac{2}{n}$, and so

$$\int_{\mathcal{M}_3} \frac{|z^2 - vw|}{1+w} {}^*dxdt \le \int_{\mathcal{M}_3} vw^*dx\,dt \le \frac{5}{n} \approx 0.$$

Finally on $\mathcal{M}_4$, $|z^2 - vw| \le \frac{vw}{n}$, and so by (2.5)

$$\int_{\mathcal{M}_4} \frac{|z^2 - vw|}{1 + w} {}^*dxdt \le \int_{\mathcal{M}} \frac{v}{n} {}^*dxdt \approx 0.$$

This completes the proof of the lemma.

For the division argument '(2.10) implies (2.11)' we need

Lemma 2.5. v is non-infinitesimal outside of a set of Loeb measure zero in $\bar{\mathcal{M}} = {}^*[0,1]\times[t_1+ 2, t_1+ 3]$.

Proof Assume the opposite. Then there is an infinitesimal $\epsilon > 0$, and a noninfinitesimal $q > 0$, such that the subset $\mathcal{S}$, where $v < \epsilon$ has *Lebesgue measure larger than q.

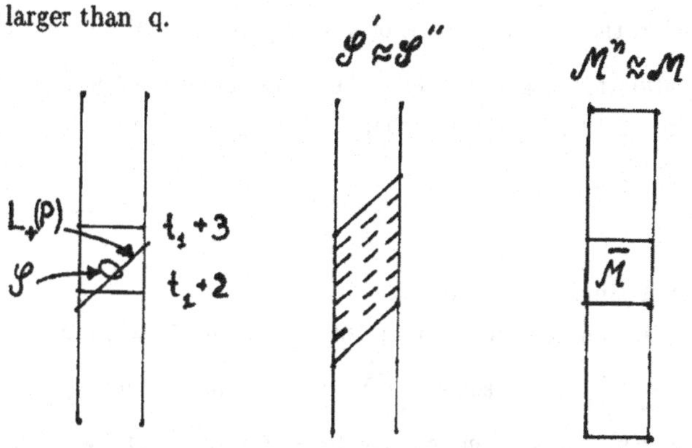

It follows from Lemma 2.4 and the equation that for some $n_1 \in {}^*N_\infty$, $\log(1+v)$ varies less than n_1^{-1} along the characteristic $L_+(P)$ for all $P \in {}^*[0,1] \times [t_1+2, t_1+3]$ outside of a subset of measure less than n_1^{-1}. Let $\mathcal{S}'$ be the union of the characteristics with this property.

Let $\mathcal{M}^n$ be the subset of $\mathcal{M}$, where $1 + w < n$. By the previous lemma for $n \in \mathbb{N}$,

$$\int_{\mathcal{M}^n} \frac{|z^2 - vw|}{1 + w} {}^*dxdt < n^{-5},$$

and so

$$\int_{\mathcal{M}^n} |z^2 - vw| {}^*dxdt < n^{-4}. \tag{2.12}$$

By overspill there is $n_2 \in {}^*\mathbb{N}_\infty$, such that (2.12) holds for $n \le n_2$. Moreover, it follows from (2.7) that the measure of the complement $\mathcal{M} \setminus \mathcal{M}^n$ is less than $C/n \log n$ with C finite.

It follows that $|z^2 - vw| < n^{-2}$ on a subset $\mathcal{S}''$ of $\mathcal{M}^n \cap \mathcal{S}'$ of measure differing from that of $\mathcal{S}'$ by less than $n^{-2} + C/n \log n$. Now $vw \le 4(\epsilon + n_1^{-1})n$ on the intersection $\mathcal{S}'''$ of $\mathcal{S}''$ with those characteristics in $\mathcal{S}'$ which intersect $\mathcal{S}$.

Choosing $n \notin {}^*N_\infty$ small enough, $(\epsilon + n_1^{-1})n \approx 0$, and so $z^2 \approx 0$ on $\mathscr{S}'''$. In particular, $z \approx 0$ outside of a set of infinitesimal measure on some characteristic $L_+(P)(\subset \mathscr{M})$. This implies that $z \approx 0$ outside of a set of infinitesimal measure in $\mathscr{M}$, if we integrate the equation for $\log(1+z)$ along the characteristic and use Lemma 2.4. And so (from $z \approx 0$ and (2.12)) we conclude that

$$vw \approx 0 \tag{2.13}$$

outside of a set of infinitesimal measure in $\mathscr{M}$.

If ${}^0v > 0$ on a noninfinitesimal subset of $\bar{\mathscr{M}}$, then ${}^0v(x,t_0) > 0$ on some noninfinitesimal set $\mathscr{M}_0 \subset {}^*[0,1]$ for some $t_1 + 2 < t_0 < t_1 + 3$. Setting $v(-x) = w(x)$, $w(-x) = v(x)$, $z(-x) = z(x)$, the problem extends to $-1 \leq x \leq 1$, and ${}^0w(x,t_0) > 0$ on $-\mathscr{M}_0$. As v and w vary infinitesimally along the backward characteristics of a noninfinitesimal subset of $\pm \mathscr{M}_0 \times \{t_0\}$, it follows that $vw \not\approx 0$ in a noninfinitesimal subset of $\mathscr{M}$ in contradiction to (2.13).

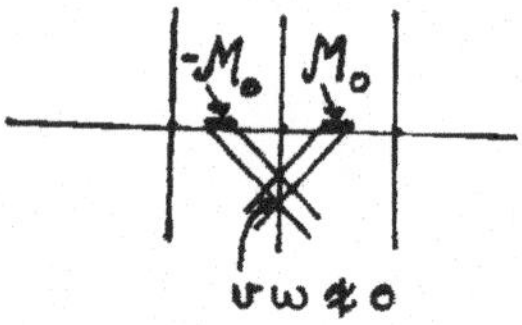

It follows that $v \approx 0$ in $\bar{\mathscr{M}}$ outside of some set of infinitesimal mesure. Analogously $w \approx 0$ in $\bar{\mathscr{M}}$ outside of a set of infinitesimal measure together with z. But the conclusion that $v, w, z \approx 0$ on a set of full Loeb measure in $\bar{\mathscr{M}}$, contradicts the mass conservation (2.4) and H-theorem (2.7). The lemma is proved.

Proof of Theorem 2.3.

By (2.12) for all P in $\bar{\mathscr{M}}$ outside of an infinitesimal set, $z^2(Q) \approx v(Q)w(Q)$ for all Q on L_- outside of an infinitesimal subset of L_-. The other two backward characteristics from Q intersect in a point P_1 along the line $\overline{P_0R}$. Again for all P

outside of an infinitesimal subset of $\bar{\mathcal{M}}$, $z^2(P_1) \approx v(P_1)\, w(P_1)$ for all Q on L_- outside of an infinitesimal subset of L_-. We may also assume that in the same sense $z(Q) \approx z(P_1)$, $u(Q) \approx w(P_1)$, $w(Q) \approx v(P)$, and that $v(Q)$ and $v(P_1)$ are finite. It follows that $vw(Q) \approx vw(P_1)$, hence that $v(Q)v(P) \approx v(Q)v(P_1)$. And so by Lemma 2.5 for all P outside of an infinitesimal subset of $\bar{\mathcal{M}}$ and for all P_1 outside of an infinitesimal subset of the corresponding line $\overline{RP_0}$, $v(P) \approx v(P_1)$.

We conclude that $v \approx C_v$ (constant) Loeb a.e. on $\bar{\mathcal{M}}$. Analogously $w \approx C_w$. By the boundary condition $C_v \approx C_w$. From $z^2 - vw \approx 0$ Loeb a.e., it follows that $z \approx v \approx w \approx C_v$ Loeb a.e. in $\bar{\mathcal{M}}$. Finally by mass conservation $C_v \approx a$. Now t_1 is arbitrary in ${}^*R^+_\infty$, and $t \to (v_t, w_t, z_t)$ is $\epsilon\delta$-continuous in ${}^*L^1$, so

$$\int |v_t - a| + |w_t - a| + |z_t - a|\, {}^*dx \approx 0, \quad t \in {}^*R^+_\infty.$$

In the standard context this means that (v_t, w_t, z_t) converges strongly in L^1 to (a,a,a) as $t \to \infty$.

Remark When working on [AI], Illner and myself first obtained this NSA proof of Theorem 2.3, and then managed to translate it back into a standard proof of convergence in measure (implying, together with an obvious weak L^1 compactness, strong L^1 convergence). But our standard proof becomes both longer and less transparent. Besides, efforts had been going on for a long time in several places to obtain this standard convergence result by purely standard techniques, whereas, once we took the NSA point of view, the present approach more or less presented itself.

References

[AI] Arkeryd, L. and Illner, R., The Broadwell model in a box, preprint 1992.

[B] Beale, J.T., Large time behaviour of a Broadwell model of a discrete velocity gas, Comm. Math. Phys. 102, 217-235 (1985).

Example 3 (On the long time behaviour of the Boltzmann equation)

This example considers the Boltzmann equation for hard and soft ($k > 2$) forces having an angular cut-off. The main result is strong L^1 convergence to Maxwellians when time tends to infinity.

The Boltzmann equation (BE) without exterior forces

$$(\partial_t + v\partial_x)f = Q(f) \tag{3.1}$$

is considered with collision operator

$$Q(f) = \int_{S^2\times R^3} (f_1'f_2' - f_1f_2)B(v_1,v_2,\theta)dndv_2,$$

where the arguments of f_1', f_2', f_1,f_2 are $(x,v_1'), (x,v_2'), (x,v_1), (x,v_2)$, and $v_1' = v_1 + n|v_1 - v_2|\cos\theta$, $v_2' = v_2 - n|v_1 - v_2|\cos\theta$, with n a unit vector in S^2. The weight function B is

$$B(v_1,v_2,\theta) = b(\theta)|v_1-v_2|^\beta \text{ with } -3<\beta\leq 1 \ (\beta = \frac{k-5}{k-1})$$

for inverse k:th power forces and $b \in L^1([0, \pi/2])$ (the angular cut-off condition). Equation (3.1) will be considered in a periodic box Λ, which after rescaling can be taken to be R^3/Z^3, with initial density distribution $f_0 = f(0)$ such that

$$f_0(1+v^2+\log f_0) \in L^1_+. \tag{3.2}$$

The following solution concepts are equivalent: the renormalized, mild, and exponential multiplier forms as in e.g. [DPL], together with the iterated integral form introduced in [A1]; f satisfies the BE in iterated integral form if $Q^\pm(f)^\#(x,v,.) \in L^1_{loc}(R_+)$ for a.e. $(x,v) \in \Lambda\times R^3$ and

$$\int_{\Lambda\times R^3} f^\#(t)\psi(t)dxdv =$$

$$= \int_{\Lambda\times R^3} f_0\psi(0)dxdv + \int_0^t ds \int_{\Lambda\times R^3} f^\#(s)\partial_s\psi \, dxdv + \tag{3.3}$$

$$+ \int_{\Lambda\times R^3} \left(\int_0^t \psi(s)Q(f)^{\#}(s)ds\right)dxdv, \quad t > 0, \quad \psi \in CL.$$

Here CL is the linear space of all functions ψ in $C^1(R_+; L^\infty(\Lambda\times R^3))$ with bounded support and with $\psi(x,v,.) \in C^1(R_+)$ for a.e. $(x,v) \in \Lambda\times R^3$. The last integral is an iterated integral. It is not required that $|\psi Q(f)^{\#}| \in L^1(\Lambda\times R^3\times[0,t])$, only that $\int_0^t \psi(s)Q(f)^{\#}(s)ds \in L^1(\Lambda\times R^3)$. Finally

$$f^{\#}(x,v,t) = f(x+vt, v,t).$$

These equivalent solutions are the ones used in the present example.

We assume that the initial value f_0 satisfies (3.2), and that the solution satisfies

$$\int_{\Lambda\times R^3} f(t)(1 + v^2 + |\log f(t)|)dxdv \leq C, \quad t \geq 0, \tag{3.4}$$

and

$$0 \leq \int_0^\infty ds \int_M ((f_1f_2)' - (f_1f_2))\log((f_1f_2)'/(f_1f_2))B\, d\mu \leq C, \tag{3.5}$$

where $M = \Lambda \times R^3 \times R^3 \times S^2$, and $d\mu = dxdv_1dv_2dn$.

The aim is now to prove strong L^1 convergence to Maxwellians when time tends to infinity.

Theorem 3.1. [A3] Given any sequence $(t_k)_N$, $t_k \nearrow \infty$, there is a subsequence $(t_{k'})$ and a global Maxwellian

$$M(v) = A \exp(-B(v-C)^2),$$

such that for $T > 0$, $f(. + t_{k'}) \to M$, strongly in $L^1(\Lambda\times R^3\times[0,T])$, and for $t > 0$, $f(.,t+t_{k'}) \to M$, strongly in $L^1(\Lambda \times R^3)$.

Proof It follows from (3.4) and the ensuing weak L^1 compactness of $(f(.+t_k))_{k\in N}$, that there is a subsequence $(t_{k'})$ such that $f(.+t_{k'}) \rightharpoonup g(.)$, weakly in $L^1(\Lambda \times R^3 \times [0,T])$ for $T > 0$. The proof will demonstrate that g equals a global Maxwellian, and discuss the strong L^1 convergence. The first part of the proof is to show that g equals a time dependent local Maxwellian

$$M(x,v,t) = a(x,t) \exp(-b(x,t)(v-c(x,t))^2).$$

For this, notice that there is a countable sequence $\phi_1, \phi_2, \ldots$ of functions with bounded support in $\Lambda \times R^3 \times R_+$, such that $g = M$ in L^1, if

$$\int g\phi_j dxdvdt = \int M\phi_j dxdvdt, \quad j \in N.$$

Also let the sequence contain $\chi_{\nu\rho}$, $v\chi_{\nu\rho}$, $v^2\chi_{\nu\rho}$, ν, $\rho \in N$, where $\chi_{\nu\rho}(v,t) = 1$ for $v^2 \leq \nu^2$, $t \leq \rho$, $\chi_{\nu\rho}(v,t) = 0$, otherwise. Set $M_k = \{\phi_1, \ldots, \phi_k\}$.

It is a consequence of the existence theory that g satisfies the BE, and that for some subsequence of $(t_{k'})$, which will from here on be denoted (t_k),

$$\int_0^k ds \int_\Lambda dx \left| \int_{R^3} f(.+t_k)\phi - g(.)\phi dv \right| < 1/k, \quad \phi \in M_k. \tag{3.6}$$

We also assume that (t_k) was so chosen that

$$\int_{t_k}^{t_k+k} ds \int ((f_1 f_2)' - (f_1 f_2)) \log((f_1 f_2)'/(f_1 f_2)) B \, dxdv_1 dv_2 dn < 1/k, \tag{3.7}$$

which is possible by (1.5).

A large part of the proof from here on relies on NSA. By transfer, in the nonstandard context (3.1) and (3.2) hold for all $k \in {}^*N$. Given $\kappa \in {}^*\mathbb{N}_\infty$, (3.1-2) implies in particular for $k = \kappa$ that

$$\int_{t_\kappa}^{t_\kappa+\kappa} ds \int ((f_1f_2)' - (f_1f_2))\log((f_1f_2)'/(f_1f_2)B\, dxdv_1dv_2dn \approx 0, \tag{3.8}$$

$$\int_0^\kappa ds \int_\Lambda dx \,|\int_{R^3} (f(t_\kappa+.)*\phi - *g*\phi)dv| \approx 0. \tag{3.9}$$

From (3.8) it follows that the integrand in the left hand side is infinitesimal; for Loeb a.a. $(x,t) \in {}^*\Lambda\times ns{}^*R_+$

$$f(x,v_1,t+t_\kappa)f(x,v_2,t+t_\kappa) \approx f(x,v_1',t+t_\kappa)f(x,v_2',t+t_\kappa) \tag{3.10}$$

for Loeb a.a. $(v_1,v_2,n) \in ns{}^*R^3\times R^3\times S^2$. Next a careful analysis of (3.10) shows that f is either Loeb a.e. infinitesimal or Loeb a.e. non-infinitesimal.

Lemma 3.2 [A2]) Let $q \in {}^*L^1_+(R^3)$ be given with

$$\int q(v)(1+v^2)*dv, \quad \int q(v)\log q(v)*dv \ \text{ finite},$$

and with

$$q(v_1)q(v_2) \approx q(v_1')q(v_2') \ \text{ for Loeb a.a. } (v_1,v_2,n) \in ns{}^*R^3\times R^3\times S^2.$$

The either $q(v) \approx 0$ for Loeb a.a. $v \in ns*R^3$, or ${}^oq(v) > 0$ for Loeb a.a. $v \in ns*R^3$.

This lemma is related to Lemma 2.5 of the previous example. For a proof the reader is referred to [A2].

It is a straight forward consequence of this lemma and (3.10) that $f(t_\kappa+.)$ is almost a local Maxwellian.

Lemma 3.3 [A2] For Loeb a.a. $(x,t) \in {}^*\Lambda\times ns{}^*R_+$, there are

$$\bar{a}(x,t),\ \bar{b}(x,t) \in R_+,\ \ \bar{c}(x,t) \in R^3,$$

such that

$$f(x,v,t+t_\kappa) \approx \bar{M}(x,v,t) \text{ for Loeb a.a. } v \in ns^*R^3,$$

with

$$\bar{M}(x,v,t) = \bar{a}(x,t)\exp(-\bar{b}(x,t)(v-\bar{c}(x,t))^2).$$

This result together with (3.9) gives for $\phi \in \bigcup_{k\in N} M_k (\subset M_\kappa)$ and $T \in R_+$

$$0 = {}^{o}\int_{*\Lambda\times[0,T]} {}^*dxdt \left| \int_{*R^3} f(x,v,t+t_\kappa)*\phi*dv - {}^*\int_{R^3} g\phi dv \right| =$$

$$= \int_{^*\Lambda\times[0,T]} Ldxdt \left| \int_{ns^*R^3} {}^{o}f(x,v,t+t_\kappa)^{o*}\phi Ldv - {}^{o*}\int_{R^3} g\phi dv \right| = \tag{3.11}$$

$$= \int Ldxdt \left| \int {}^{o}\bar{M}(x,v,t)^{o*}\phi Ldv - {}^{o*}\int g\phi dv \right|.$$

Taking $\phi = \chi_{\nu\rho} 1, \chi_{\nu\rho} v, \chi_{\nu\rho} v^2$ and letting ν tend to infinity, we get

$$\bar{a}(x,t) = {}^{o*}a(x,t),\ \bar{b}(x,t) = {}^{o*}b(x,t),$$

$$\bar{c}(x,t) = {}^{o*}c(x,t) \text{ Loeb a.a. } (x,t) \in {}^*\Lambda\times ns^*R_+,$$

where a, b, c are Lebesgue measurable functions on $\Lambda \times R_+$. It follows that ${}^{o}\bar{M} = {}^{o*}M$ with $M = a\text{ex}(-b(v-c)^2)$, i.e. $\bar{M}$ is infinitesimally close to a standard local Maxwellian.

In particular, for $\phi \in \bigcup_{k\in N} M_k$

$$\left| \int_{\Lambda\times R^3\times[0,T]} (g\phi - M\phi)dxdvdt \right| =$$

$$= \left| \int_{ns^*\Lambda^*\times R^{3*}\times^*[0,T]} ({}^{o*}g^{o*}\phi - {}^{o*}M^{o*}\phi)Ldxdvdt \right| =$$

$$= \left| \int_{{}^*\Lambda\times{}^*[0,T]} L dxdt \int_{ns{}^*R^3} L dv({}^{0*}g^{0*}\phi - {}^{0}\bar{M}^{0*}\phi) \right| \le$$

$$\le \int_{{}^*\Lambda\times{}^*[0,T]} L dxdt \left| \int_{ns{}^*R^3} L dv({}^{0*}g^{0*}\phi - {}^{0}\bar{M}^{0*}\phi \right| = 0,$$

where the last equality is (3.11). It follows that $g = M$, i.e. g is a timedependent local Maxwellian.

We have thus proved

$$f(x,v,t_\kappa + t) \approx \bar{M}(x,v,t) \approx M(x,v,t) = g(x,v,t)$$

for Loeb a.e. $(x,v,t) \in {}^*\Lambda \times ns{}^*R^3 \times ns{}^*R^+$. This implies for $T \in ns{}^*R^+$, that

$$\int_{{}^*(\Lambda\times R^3)\times[0,T]} {}^*\chi_\nu |{}^*f(t+t_\kappa) - {}^*M| {}^*dxdvdt \approx 0,$$

and that

$$\int_{{}^*\Lambda\times R^3} {}^*\chi_\nu |{}^*f(t+t_\kappa) - {}^*M| {}^*dxdv \approx 0, \text{ Loeb a.e. } t \in ns{}^*R^+. \tag{3.12}$$

By the $\epsilon\delta$-continuity of the mapping $t \to {}^*f(t)$, the relation (3.12) holds for all $t \in ns\,{}^*R^+$. From here it follows in the standard context that

$$\lim_{k\to\infty} \int_{\Lambda\times R^3\times[0,T]} |f(t+t_k) - M| dxdvdt = 0, \quad T \in R^+,$$

$$\lim_{k\to\infty} \int_{\Lambda\times R^3} |f(t+t_k) - M| dxdv = 0, \quad t \in R^+.$$

Finally, the proof that M is independent of x and t, is well-known, see e.g. [A3].

Remark i) The proof does not exclude that there are different Maxwellian limits for different sequences $(t_k)_{k\in N}$. If the energy is conserved, then the limit is unique. The same proof holds in other cases of weak L^1-limits discussed in the literature [D], implying that also those limits hold in strong L^1 sense.
ii) There does not so far seem to be any standard proof of the standard Theorem 3.1. The approach is the same as in the previous example; to make a detailed study of the consequences of the good entropy control at infinite time. The technical analysis, however, is quite different in the two problems.

References

[A1] Arkeryd, L., On the Enskog equation in two space variables, *Transp. Theory, Stat. Phys.* 15, 673-691 (1986).

[A2] Arkeryd, L., The nonlinear Boltzmann equation far from equilibrium, in *Nonstandard analysis and its applications*, ed. N. Cutland, U.K., Cambridge Univ. Press, 1988.

[A3] Arkeryd, L., On the strong L^1 trend to equilibrium for the Boltzmann equation, *Stud. Appl. Math.* to appear.

[D] Desvillettes, L., Convergence to equilibrium in large time for Boltzmann and B.G.K. equations, *Arch. Rat. Mech. Anal.* 110, 73-91 (1990).

[DPL] DiPerna, R. and Lions, P.L., Global solutions of Boltzmann's equation and the entropy inequality, *Arch. Rat. Mech. Anal.*114, 47-55 (1991).

Example 4 (The stationary Boltzmann equation)

The example discusses L^1 solutions of some stationary boundary value problems for the Boltzmann equation in a bounded slab, when the particles with a small velocity component in the slab direction have a reduced collision rate.

As a background, I will first sketch what is already known about this problem. Let $v = (\xi,\eta,\zeta) \in \mathbb{R}^3$ denote a velocity vector with x-, y- and z-components ξ, η and ζ respectively, and x the position in the slab-interval [0,a]. For two velocities $v,w \in \mathbb{R}^3$ and a collision parameter $n \in S^2$, define the collision transformation

$$J: (v,n,w) \to (v',-n,w')$$

by

$$v' = v - n(n, v-w), \; w' = w + n(n, v-w) \tag{4.1}$$

Here, $(n,v-w)$ denotes the Euclidean inner product in $\mathbb{R}^3$. J is an involution ($J^2 =$ id) and preserves momentum and energy. It is also well-known that $\|v'-w'\| = \|v-w\|$ and $|(n,v-w| \;] \; |(n,v'-w')|$, so the collision kernel $B(n,v-w)$, which in effect only depends on $\|v-w\|$ and $|(n,v-w)|$, is invariant under the action of J.

We are concerned with the steady Boltzmann equation in the slab $0 \leq x \leq a$, for $f = f(x,v)$,

$$\xi \, \partial_x f = Q(f,f) \tag{4.2}$$

with boundary conditions

$$\begin{aligned} f(0,v) &= f_0(v) \quad \text{if } \xi > 0 \\ f(a,v) &= f_a(v) \quad \text{if } \xi < 0. \end{aligned} \tag{4.3}$$

The collision operator $Q(f,f)$ is

$$Q(f,f)(x,v) = \int_{\mathbb{R}^3} \int_{S^2} B(n,v-w)[f'f'_* - ff_*]dn \, dw$$

with $f' = f(x,v')$, $f'_* = f(x,w')$ and $f_* = f(x,w)$.

Problem (4.2-3) models a kinetic layer between two walls at $x = 0$ and $x = a$, where the ingoing densities are prescribed. Our objective is to prove that this problem has a nonnegative L^1-solution.

In [ACI] Cercignani, Illner, and myself obtained measure solutions in the case of soft forces, inversely proportional to the k-th power of the distance for $3 < k < 5$, and with angular cut-off. We also had to introduce the unphysical additional truncation of a reduced interaction rate for small $|\xi|$. The proof was based on Schaefer's fixed point theorem, and depended on the flow conservation laws of the problem, but did not use any entropy arguments. In the present example the entropy derivative is used to prove that the problem has solutions x-a.e. in $L^1_{\xi 2}$, provided the ingoing boundary values have a well defined entropy flow.

To pass to the measure formulation of [ACI], choose a test function $\varphi(x,v)$, bounded and continuous, such that $\partial_x\varphi(x,v)/\xi$ is continuous and such that φ is Lipschits continuous with respect to v (with a Lipschitz constant not depending on x) and has compact support. In addition, require that

$$\varphi(0,v) = 0 \quad \text{if } \xi < 0$$

$$\varphi(a,v) = 0 \quad \text{if } \xi > 0.$$

We call such test functions "admissible". (In the case of reflexion at a, mentioned later, $\varphi(a,\cdot)$ is assumed to have the same symmetry as $f(a,\cdot)$.)

Multiply (4.2) by φ, integrate $\int_0^a \int_{\mathbb{R}^3} dv\, dx$, integrate by parts with respect to x, apply the collision transformation and use the boundary condition (4.3). The result is (in measure notation)

$$- \int_0^a \int \xi\cdot\partial_x\varphi(x,v)d\mu_x(v)dx - \int_{\xi>0} \varphi(0,v)\cdot\xi\, d\mu_0^+(v)$$

$$+ \int_{\xi<0} \varphi(a,v)\cdot\xi\, d\mu_a^-(v) \tag{4.4}$$

$$= \int_0^a \int_v \int_w \int_{S^2} (\varphi'-\varphi)B(n,v-w)dn\, d\mu_x(v)\, d\mu_x(w)dx.$$

μ_0^+ and μ_a^- are, of course, the data at $x = 0$ and $x = a$, interpreted as measures on $\xi > 0$ and $\xi < 0$ respectively.

Let the measure $dM_x(v,n,w)$ on $\mathbb{R}^3\times S^2\times\mathbb{R}^3$ be defined by

$$dM_x(v,n,w) = dn\, d\mu_x(v)d\mu_x(w). \tag{4.5}$$

Because the collision transformation J is involutive ($J^{-1} = J$), the right hand side of (4.4) can be rewritten as

$$\int_0^a \int_v \int_w \int_{S^2} \varphi(x,v)B(n,v-w)[d(M_x \circ J) - dM_x](v,n,w)dx. \tag{4.6}$$

Let $M(\mathbb{R}^3_v)$ be the cone of bounded measures on $\mathbb{R}^3_v$, endowed with the weak* topology (i.e. $\mu_n \xrightarrow[w^*]{} \mu$ if $\int \varphi \mu_n \longrightarrow \int \varphi\, d\mu$ for each continuous φ with compact support). A measure-valued function

$$\mu : [0,a] \quad x \to \mu_x \in M(\mathbb{R}^3_v)$$

is called a measure solution of (4.2-3) if $x \to \xi^2 \mu_x$ is continuous (with respect to the weak* topology on $M(\mathbb{R}^3_v)$) and (4.4) holds for all admissible test functions.

For $v = (\xi,\eta,\zeta) \in \mathbb{R}^3$ and $n \in S^2$, let n be represented by the polar angle θ (with polar axis along $v-w$) and the azimuthal angle ϕ. We assume that

$$B(n,v-w) = |v-w|^{\beta} h(\theta),$$

with h integrable on $[0,\pi]$, $\inf_{[0,\pi]} h(\theta) > 0$ (not essential) and $-1 < \beta < 0$, i.e. soft forces inversely proportional to the k-th power of the distance, $3 < k < 5$.

We introduce a factor $\chi_\epsilon(v,w,v',w')$ into the collision term which serves the necessity to reduce the collisions between particles whose velocities have small x-component. Specifically, let $\epsilon > 0$ be arbitrary but fixed and let

$$\chi_\epsilon(v,w,v',w') = \begin{cases} 1 \text{ if } \min\{|\xi|,|\xi_*|,|\xi'|,|\xi'_*|\} \geq \epsilon \\ \text{such that } \chi_\epsilon/(\xi^2 \xi_*^2) \text{ and } \chi_\epsilon/(\xi'^2 \xi_*'^2) \\ \text{are bounded otherwise (e.g. } \chi_\epsilon = 0). \end{cases}$$

Here, ξ_*, ξ' and ξ'_* denote the x- components of w, v' and w' respectively. Also assume that χ_ϵ is invariant under the collision transformation. Let

$$B_\epsilon(v,n,w) = B(n,v-w)\cdot\chi_\epsilon(v,w,v',w'),$$

and let Q_ϵ be the collision operator with B replaced by B_ϵ. A main result of this example will be that the problem

$$\xi\partial_x\mu_x = Q_\epsilon(\mu_x,\mu_x) \tag{4.7}$$

$$\mu_0|_{\{\xi>0\}} = \mu_0^+,\ \mu_a|_{\{\xi<0\}} = \mu_a^- \tag{4.8}$$

admits a $L^1_{\xi^2}$-solution in the sense that $\xi^2\mu_x = \tilde{\mu}_x$ is x-a.e. a Lebesgue absolutely continuous measure satisfying

$$-\int_0^a\int \partial_x\varphi(x,v)/\xi\, d\tilde{\mu}_x(v) - \int_{\xi>0} \varphi(0,v)\xi\, d\mu_0^+(v)$$

$$+\int_{\xi<0} \varphi(a,v)\xi\, d\mu_a^-(v) \tag{4.9}$$

$$= \int_0^a\int_v\int_w\int_n (\varphi'-\varphi)B(n,v-w)\cdot\chi_\epsilon/(\xi^2\xi_*^2)dn\, d\tilde{\mu}_x(v)d\tilde{\mu}_x(w)dx.$$

Also a crude truncation will be used. The previous parameter $\epsilon > 0$, will be kept fixed once and for all. Now choose $n \in \mathbb{N}$ and $\delta = (\log n)^{-1/8}$, and let $k_\delta = 1$ if $v^2 + w^2 \le \delta^{-2}$, $\min\{|\xi|, |\xi_*|, |\xi'|, |\xi_*'|\} > \delta$, $|v-w| > \delta$, and $k_\delta = 0$ otherwise. Let $B^\delta = B_\epsilon\cdot k_\delta$, and Q^δ be the collision operator with B replaced by B^δ, and

$$\mu\mu_*(\mu'\mu_*') \text{ by } <\mu\mu_*>_n = ff_* \wedge n\ (<\mu'\mu_*'>_n = f'f_*' \wedge n),$$

where f is the Lebesgue absolutely continuous component of μ. The problem

$$\xi\partial_x\mu_x = Q^\delta(\mu_x,\mu_x)$$

$$\mu_0|_{\xi>0} = \mu_0^+ = f_0,\ \mu_a|_{\xi<0} = \mu_a^- = f_a, \tag{4.10}$$

has a measure solution [ACI], for which $\max_{x\in[0,a]} \int \xi^2 d\mu_*(v)$ is bounded by a constant only depending on (the finite values of)

$$\int_{\xi>0} |\xi|(1+v^2)f_0 dv, \int_{\xi<0} |\xi|(1+v^2)f_a dv. \tag{4.11}$$

It is clear that μ_x is Lebesgue absolutely continuous in the present case. To prove the existence of a L^1-solution of (4.9), we shall, besides the finiteness of (4.11), also require that

$$\int_{\xi>0} \xi f_0 \log f_0 dv, \int_{\xi<0} \xi f_a \log f_a dv \tag{4.12}$$

are finite, and that $\inf f_0(v)(\inf f_a(v)) > 0$, whenever the infinimum is taken (a.e.) over a bounded set in the half-space $\xi > 0$ $(\xi < 0)$.

Remark In the case of given inflow at 0 and reflexive boundary condition at a, which can also be treated by the above approach, we have

$$<\xi f_0>_+ := \int_{\xi>0} \xi f_0 dv = \int_{\xi<0} \xi f|_{t=0} dv =: <\xi f|_{t=0}>_-,$$

and so, given $f_0(v)$ for $\xi > 0$,

$$f_0|_{t=0} = \frac{f_0}{<\xi f_0>_+} <\xi f|_{t=0}>_-, \ \xi > 0,$$

which is a diffusive boundary condition at $x = 0$. Hence also this diffusive-reflexive case is covered by the method [ACI].

Denote the solution of the boundary value problem (4.10) for Q^δ by f^n and take as test functions, suitably truncated mollifications of $\log f^n$. We obtain in the limit that

$$\int \xi f^n \log f^n|_{x=a} dv - \int \xi f^n \log f^n|_{x=0} dv - \int_0^a dx \int (Q^\delta f^n) \log f^n dv = 0 \tag{4.13}$$

Lemma 4.1. $\sup_n \int_0^a \int (Q^\delta f^n) \log f^n < \infty.$

Proof The integrand in the collision integral becomes nonnegative after the usual changes of variables. So the lemma follows if we move the known and negative pieces of the boundary integrals of (4.13) to the right, and are able to estimate the negative ones independently of n. The boundary integrals are analogous, so we discuss the one at $x = 0$, i.e.

$$\int_{A_n} \xi f^n \log f^n|_{x=0} dv,$$

where

$$A_n = \{v; \xi < 0, f^n(0,v) < 1\}.$$

From the well-known inequality

$$y \log y + v^2 y - y + e^{-v^2} \geq 0 \quad \text{for } y > 0,$$

it follows for $v \in A_n$ that

$$|\xi| f^n |\log f^n| \leq |\xi| v^2 f^n + |\xi| e^{-v^2}.$$

The integral of the right hand side can be bounded independently of n since

$$\int_{\xi<0} |\xi| v^2 f^n|_{x=0} dv + \int_{\xi>0} \xi v^2 f^n|_{x=a} = \int_{\xi>0} \xi v^2 f_0 dv + \int_{\xi<0} |\xi| v^2 f_a dv.$$

This completes the proof of the lemma.

With these preliminaries done, we are set for a NSA proof of weak x-a.e. L^1-solutions to our stationary Boltzmann equation problem (4.9). In the nonstandard context the above f^n-results hold also for a truncation at $n \in {}^*N_\infty$, i.e. $\delta = (\log n)^{-1/8} \approx 0$. Fix $n \in {}^*N_\infty$ and set $f := f^n$. We also substitute f_0 by $f_0 \wedge n$ and f_a by $f_a \wedge n$.

The mapping

$$^*[0,a] \ni x \to f(x,\cdot) \in {}^*L^1_{\xi^2}(R^3)$$

defines for each $x \in {}^*[0,a]$ a standard measure

$$\tilde{\mu}_x: \varphi \to {}^o\int {}^*\varphi\xi^2 f(x,\cdot)^*dv.$$

Evidently $\tilde{\mu}_x$ is constant as a function of x in a monad, and is therefore well defined on the standard interval $[0,a]$. We shall prove that Lebesgue a.e. in $[0,a]$, $\tilde{\mu}_x$ is Lebesgue absolutely continuous. By Lemma 4.1 and the *Fubini theorem, the integral

$$\int[\langle f'f'_*\rangle_n - \langle ff_*\rangle_n \log \frac{f'f'_*}{ff_*}\Big|_x B^{\delta *}dv\,dv_*dn \tag{4.14}$$

is finite for Loeb a.e. $x \in {}^*[0,a]$.

Lemma 4.2. If (4.14) is finite, then $\tilde{\mu}_x$ is Lebesgue absolutely continuous.

As a consequence we obtain

Theorem 4.3. The set $\mathscr{A} \subset [0,a]$ where the solution $\tilde{\mu}_x$ of (4.9) is Lebesgue absolutely continuous with respect to v, has full measure in $[0,a]$.

Proof By the discussion above, $st^{-1}\mathscr{A}$ contains a set of full Loeb measure in $^*[0,a]$. Hence $\chi_{\mathscr{A}}\circ st$ differs only by a null function from the characteristic function of $^*[0,a]$. And so $\mathscr{A}$ is Lebesgue measurable with full measure.

Proof of Lemma 4.2. We shall argue by contradiction and assume that (4.14) is finite, but $\tilde{\mu}_x$ has a Lebesgue singular component of mass $3M > 0$, say in a box $\max\{|\xi|,|\eta|,|\zeta|\} < p_1$. It follows by spillover that for some *open set $\mathcal{O}_{n_0}$ of *Lebesgue measure $1/n_0$, with $n_0 < n$, $n_0 \in {}^*\mathbb{N}_\infty$,

$$\int_{\mathcal{O}_{n_0}} \xi^2 f(x,v)^*dv > 2M.$$

On the *Lebesgue measurable subset $\mathcal{O}_1$ of $\mathcal{O}_{n_0}$, where $f > \sqrt{n_0}$, it then holds that

$$\int_{\mathcal{O}_1} \xi^2 f(x,v)^* dv > M.$$

However, this will be shown to contradict the finiteness of (4.14).

Evidently for $-1 < \beta < 0$,

$$\int_{|\xi|<\epsilon_0} d\tilde{\mu}_x(v) = \mathcal{O}(\epsilon_0^{1+\beta}) \text{ when } \epsilon_0 \to 0.$$

So it is enough for $\epsilon_0 > 0$ and standard to consider the case

$$\mathcal{O}_1 \subseteq \{v;\ \epsilon_0 < \xi < p_1, |\eta| < p_1, |\zeta| < p_1\}, \int_{\mathcal{O}_1} f(x,v)^* dv > M_1$$

(or $\epsilon_0 < -\xi < p_1$ ...). It follows from the truncations introduced, that $f \lesssim C\, n(\log n)^{1/2}$ for some finite C.

Choose p_2 and p_3 standard real such that

$$p_1 << p_3 - p_2 << p_2 < p_3.$$

Let $\mathcal{P}$ denote a *Lebesgue measurable subset of noninfinitesimal measure of

$$p_2 < \xi < p_3,\ |\eta| < p_1,\ |\zeta| < p_1,$$

where $f(x,\cdot)$ is bounded from above by a finite constant, and from below by a noninfinitesimal positive constant.

Consider the subset $\mathcal{S}$ of $\mathcal{O}_1 \times \mathcal{P} \times {}^*S^2$ where

$$<f'f'_*>_n < (n_0 \wedge <ff_*>_n)^{1/2}.$$

Then

$$\int_{\mathcal{S}} <ff_*>_n \log ff_* B^{\delta *} dv dv_* dn$$

is finite. Consider the subset $\mathcal{S}'$ of $\mathcal{S}$ where $<ff_*>_n = n$. Evidently

$$\int_{\mathscr{S}'} f(\log f)^{1/2*} dv dv_* dn$$

is finite. Also

$$\int_{\mathscr{S}-\mathscr{S}'} ff_* \log ff_* B^{\delta *} dv dv_* dn$$

is finite, hence also

$$\int_{\mathscr{S}-\mathscr{S}'} f \log f \,{}^* dv dv_* dn.$$

Then the corresponding integral over $\mathcal{O}_1 \times \mathscr{P} \times {}^*S^2 - \mathscr{S}$ is infinite, since $\int_{\mathcal{O}_1} f(x,v)dv > M_1$, and $f(x,v) > \sqrt{n_0}$ on $\mathcal{O}_1$. This implies (by spillover) that for some $j \in {}^*N_\infty$ and for all $v \in \mathcal{O}_1$ outside of a subset $\tilde{\mathcal{O}}$ with

$$\int_{\tilde{\mathcal{O}}} f(x,v)^* dv < \frac{1}{j},$$

it holds for each $v_* \in \mathscr{P}$ outside of a *Lebesgue measurable subset of measure less than $1/j$, that outside of a $\frac{1}{j}$ - fraction of the sphere with poles at v and v_*

$$\langle f'f'_* \rangle_n \geq (n_0 \wedge \langle ff_* \rangle_n)^{1/2}. \tag{4.15}$$

In particular given $v \in \mathcal{O}_1 - \tilde{\mathcal{O}}$, we can choose v_* on a set I of non-infinitesimal measure of a line segment in $\mathscr{P}$ with $p_2 < \xi_* < p_3$, so that (4.15) holds Loeb a.e. with respect to $(v_*,n) \in I \times {}^*S^2$. Clearly, this subset of $I \times {}^*S^2$ is mapped into a non-infinitesimal volume of $v'(v'_*)$ between the extremal spheres under this construction.

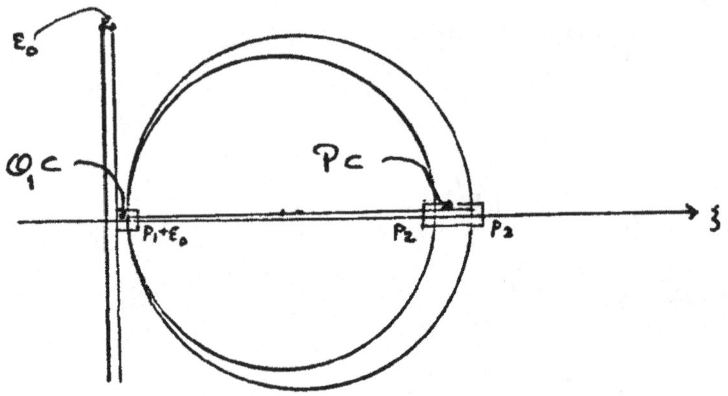

Since the product $f(x,v')f(x,v'_*)$ is infinite in this case, at least one of $f(x,v')$ and $f(x,v'_*)$ is infinite on a non-infinitesimal set of v' or v'_* in $ns^*\mathbb{R}^3$. That, however, contradicts the finiteness of

$$\int_{*R^3} \xi^2 f(x,v)^* dv,$$

and the lemma follows.

Remark i) Theorem 4.3 holds also when the condition of given incoming mass flow at $x = a$, is substituted by reflexion at $x = a$, as well as for the case of a diffusive boundary condition at $x = 0$ and reflexion at $x = a$. For the case of given incoming mass flow at $x = 0$, reflexion at $x = a$ and a infinite, some information about the Milne problem can be obtained by combining the above approach with Lemma 3.3.
ii) So far there is no standard proof of the main result, Theorem 4.3. The key lemma, Lemma 4.2, uses a nonstandard nice representation (here a L^1-function) of what, from the standard point of view, is a cruder object (here a measure) to obtain new information (from the entropy derivative, which here is well defined only in the nonstandard setup).

References

[ACI] Arkeryd, L., Cercignani, C. and Illner, R., Measure solution of the steady Boltzmann equation in a slab, *Comm. Math. Phys.* 142, 285–296 (1991).

Example 5. The final example considers a gas of infinitely many particles in the nonstandard context, evolving in $^*L^1$ according to a (stochastic) Liouville equation. The corresponding set of j-particle densities f_j, which satisfy the so-called BBGKY hierarchy of equations, can for finite j be interpreted as standard distributions through $\varphi \to {}^o\!\int f_j{}^*\varphi$. With L^1 initial data, these distributions turn out to be L^1-solutions of the Boltzmann hierarchy (BH) which factorize if the initial data do. This is the expected behaviour of a gas in the Boltzmann Grad limit from the point of view of statistical physics. Mathematically, however, one would like to obtain more, namely suitably strong convergence results *within* the standard L^1-context, when the (finite) number of particles tends to infinity.

This example is an observation originating from ongoing research together with S. Caprino and M. Pulvirenti [ACP], and centered on certain Boltzmann gases with one space- and three velocity-dimensions. It will here be described for the technically less complicated Lebowitz model, a very special gas of sticks. Let me start by giving a background sketch of the Lebowitz gas of sticks [FL] in the (x,y)-plane R^2. Each stick is oriented along the y-axis, and moves in a straight line with uniform velocity until it collides with another stick side to side. Then the two sticks exchange their velocity component v_x in the x-direction.

On the particle level for a gas of N sticks, the evolution can e.g. be given by the time evolution of a density f^N through the Liouville equation. We will use a stochastic version of the Liouville equation and take f^N as a probability density, symmetric under particle exchange. From the Liouville equation, the BBGKY hierarchy is obtained by integrating away all but the first j particles. At a cruder scaling the evolution can be described by gas kinetics. The *simplifying property* of the stick model on the particle level, is that it partly decouples the equations of the hierarchy if the densities are y-independent. On the kinetic level the Boltzmann equation for the sticks can similarly be decoupled into linear equations. We shall rely on that simplifying property and so drop the y-dependence spacewise, only considering phase points $z^N = (z_1, z_2, \ldots, z_n)$ with $z_j = (x_j, v_{xj}, v_{yj}) = (x_j, v_j) \in R^3$.

As a first step let us discuss the Liouville equation. Introduce the notations

$$f_t^{N\#}(z^N) = f_t^N(x_1 + tv_{x1}, v_{x1}, v_{y1}, \ldots, x_N + tv_{xN}, v_{xN}, v_{yN}),$$

$$J_{ik}(x^N, v^N, t) = (x_1, v_1, \dots, x_{i-1}, v_{i-1}, x_i + t(v_{xi} - v_{xk}), v_i', x_{i+1}, v_{i+1}, \dots$$

$$x_{k-1}, v_{k-1}, x_k + t(v_{xk} - v_{xi}), v_k', x_{k+1}, v_{k+1}, \dots x_N, v_N),$$

where $v_i' = (v_{xk}, v_{yi})$, $v_k' = (v_{xi}, v_{yk})$, and

$$(L^{+\#}(t) - L^{-\#}(t))f(x^N, v^N) = L^{\#}(t)f(x^N, v^N) =$$

$$= \sum_{i<k} |v_{xi} - v_{xk}| [f o J_{ik}(x^N, v^N, t) - f(x^N, v^N)] \delta(x_i - x_k + t(v_{xi} - v_{xk}))$$

with δ the Dirac measure. We shall write $L := L^{\#}(0)$, and consider the stochastic Liouville equation given by

$$D_t f_t^{N\#} = \epsilon (L f_t^N)^{\#} \quad (\epsilon = \tfrac{\lambda}{N}) \text{ with } f_{t=0}^N = f_0^N \in L_+^1 . \tag{5.1}$$

Here $\int f_0^N(z^N) dz^N = 1$ (normalization), and $f_0^N(Pz^N) = f_0^N(z^N)$ (particle symmetry) for any permutation P of $1, \dots, N$.

Formally the expansion

$$f_t^{N\#}(x^N, v^N) := \sum_{n=0}^{\infty} \epsilon^n \int_0^t dt_1 \dots \int_0^{t_{n-1}} dt_n L^{\#}(t_1) \dots L^{\#}(t_n) f_0(x^N, v^N) \tag{5.2}$$

satisfies the above equation in mild form

$$f_t^{N\#} = f_0 + \epsilon \int_0^t (L f_s^N)^{\#} ds.$$

For $N = 2$ this gives for $t > 0$

$$f_t^{\#}(z^2) = f_0 \text{ if } t_0 := (x_1 - x_2)/(v_{x2} - v_{x1}) \le 0.$$

If $t_0 > 0$ then for $0 < \epsilon < 1$

$$f_t^{\#}(z^2) = f_0(z^2), \quad t < t_0,$$

$$f_t^{\#}(z^2) = \frac{1}{2}(1 + e^{-2\epsilon})f_0(z^2) + \frac{1}{2}(1 - e^{-2\epsilon})f_0(x_1, v_{x1}, v_{y2}, x_2, v_{x2}, v_{y1}), \; t > t_0.$$

In this case positivity and L^1 norm are conserved.

If $f_0 \in L^\infty$ and has compact support in $|x_j| \le a$, $|v_{x_j}| \le a$, $|v_{yj}| \le a$, $j = 1, \ldots, N$, then the above formal reasoning is strict in strong L^1 sense, since $(s_i = \pm 1)$

$$\begin{aligned} &\int dx^N dv^N \epsilon^n \int_0^t dt_1 \ldots \int_0^{t_{n-1}} dt_n L^{s_1\#}(t_1) \ldots L^{s_n\#}(t_n) f_0^N \le \\ &\le \|f_0\|_\infty (a^2(a + 2ta)^N (N\lambda)^n. \end{aligned} \tag{5.3}$$

If the Dirac measure in the Liouville operator L is replaced by a symmetrically mollified version, then the new measure-factors in L become pointwise bounded. The corresponding Liouville operator will be denoted by $L_{\mathcal{M}}$. In this case the right hand side of (5.3) can be improved to

$$\|f_0\|_\infty (a^2(a + 2ta))^N (CN\lambda t)^n / n!,$$

which increases the radius of convergence from $\sim 1/N\lambda$ to ∞. One can show that the solution of the mollified equation with N fixed, converges strongly in L^1 to the solution (5.2) for L, when the mollified Dirac measure tends to the Dirac measure. In the mollified case the solution is unique, and can easily (using the exponential form) be shown to be non-negative.

It is an immediate consequence of the definition of $L_{\mathcal{M}}$, that

Lemma 5.1. If f is particle symmetric, then $L_{\mathcal{M}}^{\#} f$ is particle symmetric.

The present analysis requires an N-independent L^1-estimate of each contracted term in the mollified version of $f_t^{N\#}$.

Lemma 5.2. Let $h \in N$. Then

$$\epsilon^h \int_{R^{3m}} dz^m \Big| \int_{R^{3(N-m)}} dz_{m+1} \dots dz_N \int_0^t dt_1 \dots \int_0^{t_{h-1}} L^{\#}_{\mathscr{M}}(t_1) \dots L^{\#}_{\mathscr{M}}(t_h) f \Big| \le$$

$$\begin{bmatrix} m+h-1 \\ h \end{bmatrix} (2\lambda a C t)^h \int_{R^{3N}} |f| dz^N.$$

Proof. $$\epsilon \int_{R^{3m}} dz^m \Big| \int_{R^{3(N-m)}} dz_{m+1} \dots dz_N \int_0^t dt_1 L^{\#}_{\mathscr{M}}(t_1) f \Big| \le$$

$$\left[\frac{1}{N}(N-m)2m + \frac{m(m-1)}{N}\right] (2\lambda a C t) \int_{R^{3(m+1)}} dz^{m+1} \Big| \int_{R^{3(N-m-1)}} f dz_{m+2} dz_N \Big| \le$$

$$m(2\lambda a C t) \int_{R^{3(m+1)}} dz^{m+1} \Big| \int_{R^{3(N-m-1)}} f dz_{m+2} \dots dz_N \Big|.$$

Using iteration and noting that, by Lemma 5.1, particle symmetry is preserved, this proves the lemma for $h \le N-m$. When $h > N-m$

$$\int_{R^{3N}} \Big| \int_0^t \epsilon L^{\#}(t_1) f dt \Big| dz^N \le \frac{N(N-1)}{N} (2\lambda a C t) \int_{R^{3N}} |f| dz^N,$$

and so

$$\epsilon^h \int_{R^{3m}} dz^m \Big| \int_{R^{3(N-m)}} dz_{m+1} \dots dz_N \int_0^t dt_1 \dots \int_0^{t_{h-1}} dt_h L^{\#}(t_1) \dots L^{\#}(t_h) f \Big| \le$$

$$\le \frac{(N-1)! N^{h-N+m}}{(m-1)! h!} (2\lambda a C t)^h \int_{R^{3N}} |f| dz^N.$$

The L^1 convergence of the BBGKY hierarchy to the Boltzmann hierarchy follows using Lemma 5.2. The following notations will be convenient in our analysis of that convergence. Take

$$f_k^N = \int f^N dz_{k+1} \ldots dz_N,$$

and let

$$(m+1 \,|\, m+i) = (m+1, m+2, \ldots, m+i), \; \alpha = (\alpha_1, \ldots, \alpha_i), \; \beta = (\beta_1, \ldots, \beta_i)$$

denote multiindices with

$$\beta < \alpha \quad \text{if} \;\; \beta_1 < \alpha_1, \ldots, \beta_i < \alpha_i.$$

For $f = f(z^k)$ set

$$\Delta_{ik}^t f(z^{k-1}) = \int_0^t ds \int_{R^3} dz_k \, |v_{xk} - v_{xi}| \, [f \circ J_{ik}(x^k, v^k, s) - f(x^k, v^k)] \, \delta(x_i + v_{xi} t - x_k - v_{xk} t),$$

and define for $T_i = (t, t_1, \ldots t_{i-1})$, $\alpha = (\alpha_1, \ldots, \alpha_i) < \beta = (\beta_1, \ldots, \beta_i)$;

$$\Delta_{\beta\alpha}^{T_i} = \Delta_{\beta_1 \alpha_1}^t \;\ldots\; \Delta_{\beta_i \alpha_i}^{t_{i-1}}.$$

We have by integration of $(5.1)_{\mathcal{M}}$ (i.e. (5.1) with $L_{\mathcal{M}}$ instead of L)

$$D_t f_{k\,t}^{N\#} = \epsilon (L_{\mathcal{M}} f_{kt}^N)^{\#} + \epsilon (N-k) (Q_{k,k+1}^{\mathcal{M}} f_{k+1,t}^N)^{\#}, \tag{5.4}$$

where

$$(Q_{k,k+1}^{\mathcal{M}} f_{k+1}^N)^{\#}(z^k, t) = \sum_{i=1}^{k} \int dz_{k+1} \, |v_{xi} - v_{xk+1}| \, [f_{k+1}^{N\#} \circ$$

$$J_{i,k+1}(z^{k+1}, t) - f_{k+1}^{N\#}(z^{k+1}, t)] \, \varphi_{\mathcal{M}}(x_i + v_{xi} t - x_{k+1} - v_{xk+1} t),$$

and $\varphi_{\mathcal{M}}$ is the mollified Dirac measure.

Assume that $f_k^N \to f^k$ in L^1-norm as $N \to \infty$, $k = 1,2,\dots$. Then there holds the following termwise contraction limit for (5.2).

Lemma 5.3.

$$N^{-h} \int_{R^{3(N-m)}} dz_{m+1}\dots dz_N \int_0^t dt_1 \dots \int_0^{t_{h-1}} dt_h L^{\#}_{\mathcal{M}}(t_1)\dots L^{\#}_{\mathcal{M}}(t_h) f^N \to$$

$$\sum_{\beta<(m+1|m+h)} {}_{\mathcal{M}}\Delta^{T_h}_{\beta\,,\,(m+1|m+h)} f^{m+h}(z^m),$$

strongly in L^1 as $N \to \infty$. Here δ is replaced by $\varphi_{\mathcal{M}}$ in ${}_{\mathcal{M}}\Delta$.

Proof. For $h = 1$ by the proof of Lemma 5.2

$$\frac{1}{N} \int_{R^{3(N-m)}} dz_{m+1}\dots dz_N \int_0^t dt_1 L^{\#}_{\mathcal{M}}(t_1) f^N \to \sum_{i=1}^m {}_{\mathcal{M}}\Delta^t_{i,m+1} f^{m+1}(z^m).$$

By induction over h the lemma follows.

Using Lemma 5.3 the convergence of the BBGKY hierarchy to the BH follows.

Lemma 5.4.

$$\lim_{N\to\infty} \int_{R^{3(N-m)}} dz_{m+1}\dots dz_N \sum_{k=0}^{\infty} \epsilon^k \int_0^t dt_1 \dots \int_0^{t_{k-1}} dt_k L^{\#}_{\mathcal{M}}(t_1)\dots L^{\#}_{\mathcal{M}}(t_k) f^N$$

$$= \sum_{k=0}^{\infty} \lambda^k \sum_{\beta<(m+1|m+k)} {}_{\mathcal{M}}\Delta^{T_k}_{\beta,(m+1|m+k)} f^{m+k}(z^m), \tag{5.5}$$

strongly in L^1.

Proof. Given N we know that the series solution of (5.1) is absolutely convergent in L^1 for $0 \le t < \infty$. So, given N, by dominated convergence the order of integration and summation can be reversed in the left hand side of (5.5). By Lemma 5.2

$$\sum_{k=0}^{\infty} \int_{R^{3m}} \left| \int_{R^{3(N-m)}} \epsilon^k \int_0^t dt_1 \ldots \int_0^{t_{k-1}} dt_k L^{\#}_{\mathscr{M}}(t_1)\ldots L^{\#}_{\mathscr{M}}(t_k) f^N \right| \leq$$

$$\leq \sum_0^{\infty} \begin{bmatrix} m+k-1 \\ k \end{bmatrix} (2\lambda aCt)^k = D_s^{m-1} s^{m-1} e^s \Big|_{s=2\lambda aCt}.$$

Here the right hand side is independent of N. This together with Lemma 5.3 implies that the sum of the contracted terms converges strongly in L^1 to the right hand side of (5.5).

Using e.g. an approximation with alternatingly the flow and the collisions turned on, together with an algebraic analysis of the right hand side in (5.5), one can prove the following factorization result.

Lemma 5.5. If $f^k(z^k) = \prod_{\nu=1}^{k} f_0(z_\nu)$, then

$$\sum_{k=0}^{\infty} \lambda^k \sum_{\beta<(m+1|m+k)} \Delta^{T_k}_{\beta,(m+1|m+k)} f^{m+k}(z^m) =$$

$$= \prod_{\nu=1}^{m} \left\{ \sum_{k=0}^{\infty} \lambda^k \sum_{\beta<(2|1+k)} \Delta^{T_k}_{\beta,(2|1+k)} f^{1+k}(z_\nu) \right\}.$$

Let $f_0 \in L^1_+(R^3)$ be given with $\int f_0(z)dz = 1$ and with f_0 belonging to a bounded subset with respect to the integral

$$\int (1 + z^2 + |\log f_0(z)|) f_0(z) dz.$$

For $t > 0$ the L^1-solution f_t of the Boltzmann equation for the Lebowitz sticks depends continuously in the L^1 norm on f_0. In particular this holds (when $a \to \infty$) for the solutions with initial values

$$\bar{f}_a = f_a / \int f_a(z)dz,$$

where

$f_a = f_0 \wedge a$ for $\max(|x|, |v_x|, |v_y|) \leq a$,

$f_a = 0$ otherwise.

Also for $t > 0$ the (unique) factorized L^1-solution of the mollified BH with initial value $f_0^k(z^k) = \prod_{\nu=1}^{k} \bar{f}_a(z_\nu)$, converges strongly in L^1 to the factorized solution of the BH (without mollification) for the sticks, when $\varphi_{\mathcal{M}}$ converges to δ. (It is enough to study the lowest level, i.e. the Boltzmann equation, e.g. using a contraction mapping argument in the norm

$$\int \sup_{0 \leq t \leq T} |f_t^{\#}(z)| dz,$$

cf. the approach in [A].)

The above analysis can by transfer be carried out in the nonstandard context. Together with an overspill argument, this easily leads to the result mentioned at the beginning of the example. For simplicity we only discuss the case of factorized initial data. Evidently, given $N \in {}^*\mathbb{N}_\infty$, for infinitesimally mollified Dirac measures in $L_{\mathcal{M}}$, and for infinite $a \in {}^*R_\infty^+$, by transfer there is a unique solution f_t^N to the Liouville equation $(5.1)_{\mathcal{M}}$ with initial value $f_0^N = \prod_1^N \bar{f}_a(z_j)$.

By overspill for *some* infinitesimally mollified $L_{\mathcal{M}}$ and *some* infinite $a \in {}^*R_\infty^+$, the following result holds.

Theorem 5.6. For all finite k the k-particle contractions f_{kt}^N are for finite time t infinitesimally close in the ${}^*L^1$-norm to the unique factorized solutions of the BH for the (true) Lebowitz gas. For $\varphi(z^k)$ a standard test function, the standard part of each term in the k-th equation of the weak BBGKY hierarchy, i.e. each term of

$$\int f_{k\,t}^{N\#}(z^k)^* \varphi_t(z^k)^* dz^k - \int_0^t \int f_{k\,s}^{N\#*} \partial_s \varphi_s{}^* dz^k ds =$$

$$\int \prod_1^k \bar{f}_a(z_j)^* \varphi_0(z^k)^* dz^k + \frac{\lambda}{N} \int_0^t \int (L_{\mathcal{M}} f_{ks}^N)^{\#} \, {}^*\varphi_s{}^* dz^k ds$$

$$+ \frac{\lambda(N-k)}{N} \int_0^t \int (Q^{\mathcal{M}}_{k,k+1} f^N_{k+1,s})^{\#} * \varphi_s * dz^k ds,$$

equals the corresponding term in the corresponding equation from the weak BH, i.e.

$$\int \prod_1^k f_t^{\#}(z_j)\, \varphi_t(z^k) dz^k - \int_0^t \int \prod_1^k f_s^{\#}(z_j) \partial_s \varphi_s dz^k ds =$$

$$\int \prod_1^k f_0(z_j) \varphi_0(z^k) dz^k + \lambda \int_0^t \int (Q_{k,k+1} \prod_1^k f_s)^{\#} \varphi\, dz^k ds$$

(the $L_{\mathcal{M}}$-term gives zero).

Remark. In the previous examples, NSA was used for the preliminary analysis, and/or as a technical tool to obtain standard results. In this example, however, the (hard) computations are all in the standard context, and lead, via transfer and overspill, to a description, Theorem 5.6, of the Lebowitz gas, which in the NSA context is an N-particle system evolving according to a Liouville equation, and in the standard context a kinetic gas evolving according to the Boltzmann hierarchy.

References

[A] Arkeryd, L., Existence theorems for certain kinetic equations and large data, *Arch. Rat. Mech. Anal.* 103, 139-149 (1988).

[ACP] Arkeryd, L., Caprino S., Pulvirenti M., In preparation.

[FL] Frisch, H.L., Lebowitz, J.L., Model of nonequilibrium ensemble, *Phys. Rev.* 107, 917-923 (1957).

GLOBAL SOLUTIONS OF KINETIC MODELS AND RELATED QUESTIONS.

P.L. Lions

Ceremade
Université Paris-Dauphine
Place de Lattre de Tassigny
75775 Paris Cedex 16

Summary

I. Introduction.
II. Presentation of models and general observations.
III. Existence and convergence results.
IV. Velocity averaging.
V. Generalized flows.

References.

I. Introduction.

We shall review here some recent progress (and indicate some new results as well) on various *kinetic equations* which include the well-known *Boltzmann equation* and *Vlasov models* like, for instance, Vlasov-Poisson or Vlasov-Maxwell systems. Before presenting these models in section II, we would like to recall first a few of the basic physical notions underlying these models.

First of all, kinetic models (and their applications to various fields of Physics) are a branch of *Statistical Physics*. They arise in a large number of physical situations like, for instance, in the study of the dynamics of electrons and ions in plasmas and in lasers, in the study of the dynamics of nucleons in Nuclear Physics, in the modelling of semi-conductors, in the modelling of hypersonic flights or of the reentry of aircrafts in a rarefied atmosphere, in the study of the formation (and stability) of planetary rings or even in the study of the formation of galaxies. It is worth emphasizing that such a list, by no means exhaustive, involves different physical interactions on different scales. Of course, each of these physical situations reveals specific mathematical questions that we will not address here. Instead, we shall be mainly concerned with some of the basic mathematical issues that are raised in all of these applications. More precisely, these models are evolution type problems and we will investigate the structure of the Cauchy problems associated to these evolution equations.

In spite of these extremely different physical backgrounds, the main principle underlying these models can be summarized as follows. Let us suppose we want to study the evolution of a large number of particles that we take to be identical in order to simplify the presentation. The word particle is used with a vague meaning since these particles can be classical point particles or quantum particles and in the physical applications listed above the "particles" may be electrons, nucleons, ions, gas molecules, rocks or even stars !

We next observe that, as the number of particles increases, following the dynamics of each particle becomes soon untractable and one then wishes to look instead for a statistical description of this evolution. In order to be more precise, let us assume that we deal with classical point-particles. In that case, the unknowns which would have been otherwise the position and velocities of each particle "reduce" to a function f of (x, ξ, t) $(x, \xi \in \mathbf{R}^N, t \geq 0)$ which is the density of particles at position x, time t and with velocity ξ. Of course, f is nonnegative. In that situation, kinetic models often take (but not always) the following form

$$\frac{\partial f}{\partial t} + \xi \cdot \nabla_x f + F \cdot \nabla_\xi f = C \qquad \text{in } \mathbf{R}^{2N} \times (0, \infty) . \tag{1}$$

Here and everywhere below, $\nabla_x f$ and $\nabla_\xi f$ denote respectively the gradient of f with respect to x and ξ, and we use $a \cdot b$ or (a, b) indifferently for the scalar product in $\mathbf{R}^N$ of two vectors a and b.

In the equation (1), the term F represents a *force* field acting on the particles and it will in fact depend on the solution f itself as we show in the examples presented in the next section. In that case the term $F \cdot \nabla_\xi f$ becomes nonlinear (it is in fact typically quadratic and non-local). Similarly, C stands for a term which takes into account the possible *collisions* between particles. Again, C is a nonlinear term which is typically quadratic. Examples like the famous Boltzmann and Vlasov models will be given in section II.

Let us now briefly describe the numerous mathematical problems raised by these models. As we already mentioned above, we shall concentrate on the main mathematical issues concerning these Cauchy problems, forgetting many important questions like boundary conditions or specific questions of interest for one or several of the physical applications listed above.

The first category of problems is the study of the Cauchy problems for these models, with the usual questions regarding existence, uniqueness, regularity, approximation and numerical analysis, special solutions, steady states, stability, long-time behaviour...

But these is much more to be studied and understood. In fact, it seems fair to say that the Boltzmann equation is also well-known because of its formal relationship with other famous mathematical physics models like hydrodynamical models (compressible Euler equations, i.e. gas dynamics systems, incompressible Navier-Stokes or Stokes or Euler equations...). It is worth recalling Hilbert goal to solve Boltzmann equation *and* to recover from it the Fluid Dynamics equations. Therefore, an important category of mathematical problems concerns the systematic study of the numerous links between kinetic models and other models in Physics. We just mentioned hydrodynamical limits and the link with Fluid Dynamics but one has to add to that theme the derivation of MHD equations, combustion models, reaction-diffusion systems and in fact rather general hyperbolic systems of conservation laws.

Other connections with various regimes and models of Physics exist and lead to relevant mathematical issues. In particular, other limits involve the derivation of kinetic models from limits of classical Newtonian mechanics models when the number of particles goes to infinity and the study of statistical solutions or hierarchies of equations and the propagation of chaos. Another example is given by the derivation of kinetic models from Quantum models via Wigner transforms and semi-classical limits...

Of course, these two categories of problems are not clearly separated and one may expect that progress in the analysis of the Cauchy problems leads to progress in these various limits. And in fact, even if our main emphasis will be in the study of the Cauchy problems, we want to mention that semiclassical limits ($\hbar \to 0$) can be studied by convenient adaptations of the arguments we discuss here (see P.L. Lions and T. Paul **[29]**; P. Gérard, P.L. Lions and T. Paul **[19]**) and that some progress on hydrodynamical limits (and related questions) can be found in C. Bardos, F. Golse and D. Levermore **[3]** ; P.L. Lions, B. Perthame and E. Tadmor **[31]**. And we emphasize the fact that the tools introduced for the analysis of the Cauchy problems allow to make significant progress, we hope, on such limits.

To explain the type of results on the Cauchy problems associated with kinetic models we will be discussing, it is worth explaining the three types of results that have been obtained on the Boltzmann equation, for instance, restricting our attention to the basic existence and uniqueness questions :

- smooth and unique solutions in the small that is locally in time or globally with smallness conditions : for the Boltzmann equation alone, many important works have been given in that direction and we can quote only a few of them (complete lists of references can be found in the references given in the bibliography here) like H. Grad [**23**], C. Cercignani [**7**], R. Illner and M. Shinbrot [**25**], T. Nishida and K. Imai [**33**], S. Ukaï [**37**], K. Hamdache [**24**]... References can be found in R.J. DiPerna and P.L. Lions [**11**], C. Cercignani [**7**]...

- existence and uniqueness of special solutions : a famous example is given by the study of space-homogeneous solutions, i.e. x-independent solutions of Boltzmann equation, study initiated by the work of T. Carleman [**6**] - see [**11**],[**7**] for more references in that direction...

- global existence of weak solutions.

This last category of results has been obtained in a series of works for most kinetic models by R.J. DiPerna and the author [**11**]-[**15**] (see also the survey [**16**] and [**27**]), with some recent developments discussed also below.

We present these results in section III after recalling in section II the precise forms of the equations we are studying i.e. the precise forms of F and C in (1). Also in section III we give the main convergence results (or stability results) associated to these existence results. A word of explanation is necessary there : indeed, as it is often the case for evolutionary nonlinear partial differential equations, some a priori estimates are available (they are recalled in section II) and they follow from conserved quantities (that are in turn associated to symmetries and invariances...). The heart of existence proofs for global weak solutions is often the ability to pass to the limit in the equations with the help of those a priori estimates. The actual existence follows then from some ad hoc approximation procedure that yields the desired sequences of (almost) solutions on which we can perform the passage to the limit. And the stability results we mentioned above are precisely what we need for such passages to the limit (that are in general delicate since, in general, only weak convergences are possible and the equations incorporate nonlinear terms).

This is why we shall insist here on the proofs of such stability results and elements are given in these notes. Indeed, in section IV and V we recall briefly the main tools that are being used in these proofs namely velocity averaging lemma (section IV) and the construction of generalized ODE flows (section V).

The global solutions we build are weak ones. Outstanding open problems are the issues of the regularity of solutions and their uniqueness. The only answer to these basic questions we can offer concerns the collisionless case (i.e. where $C = 0$) in which case we deal with a "pure" Vlasov system. This case is treated in P.L. Lions and B. Perthame [30] (see also for less general results K. Pfaffelmoser [34], R. Glassey [20], J. Schaeffer [35]).

II. Presentation of models and general observations.

We begin with *collision-less models* i.e. $C = 0$ in (1). The typical model is the so-called *Vlasov-model*. We just have to specify the force F which in Vlasov models is a self-consistent (or mean-field) force i.e. F depends upon f in general in a linear, nonlocal way. A first class of models concerns Newtonian point particled in which case we have

$$F = -\nabla_x V \quad , \quad V = V_0 * \rho \quad , \quad \rho = \int_{\mathbb{R}^N} f \, d\xi \tag{2}$$

where V_0 is a prescribed even function on $\mathbb{R}^N$ corresponding to an interaction potential. We shall call *Vlasov system* ((V) in short) the system obtained by combining (1) with $C = 0$ with (2). Probably, the most interesting example arising in Physics is the so-called *Vlasov-Poisson system* where $N = 3$ and $V_0 = \pm\frac{1}{|x|}$; and both signs are relevant since $+$ corresponds to charged particles and to electrostatics while $-$ corresponds to massive particles and classical gravitation interactions.

Another interesting model is the so-called *Vlasov-Maxwell model* where the force F is given by the Lorentz force (and $N = 3$)

$$F = E + \frac{1}{c} \xi \times B \tag{3}$$

and (E, B) the electromagnetic field satisfies Maxwell's equations

$$\begin{cases} \dfrac{\partial E}{\partial t} - c \text{ curl } B = -j \quad , \quad \text{div } B = 0 \\ \dfrac{\partial B}{\partial t} + c \text{ curl } E = 0 \quad , \quad \text{div } B = \rho \end{cases} \tag{4}$$

with

$$\rho = \int_{\mathbb{R}^N} f \, d\xi \qquad j_k = \int_{\mathbb{R}^N} f \, \xi_k \, d\xi \qquad (k = 1, 2, 3) . \tag{5}$$

Here, c represents the speed of light. In all the following, we denote by (VM) the Vlasov-Maxwell system composed of (1) (with $C = 0$), (3),(4) and (5).

Let us recall that in all these collision-less models, (1) simply means that f is constant along particle paths ($\dot{x} = \xi$, $\dot{\xi} = F$) and that the vector-field (ξ, F) is always conservative i.e.

$$\operatorname{div}_{(x,\xi)}(\xi, F) = 0 . \tag{6}$$

We next present collision models i.e. we give a few important examples of terms C. We begin with the classical model due to L. Boltzmann [**5**] (see also J.C. Maxwell [**32**])

$$C = Q(f,f) = \iint_{\mathbb{R}^N \times S^{N-1}} d\xi_* \, d\omega \, B(\xi - \xi_*, \omega) \{ f' f'_* - f f_* \} . \tag{7}$$

Here and below, we use the notation : $f_* = f(x, \xi_*, t)$, $f' = f(x, \xi', t)$, $f'_* = f(x, \xi'_*, t)$ and $\xi' = \xi - (\xi - \xi_*, \omega)\omega$, $\xi'_* = \xi_* + (\xi - \xi_*, \omega)\omega$.

We shall always assume that $B(z,\omega)$ depends only on $|z|$ and $|(z,\omega)|$, $B \geq 0$ and B satisfies

$$\begin{cases} B \in L^1_{\text{loc}}(\mathbb{R}^N \times S^{N-1}) \quad ; \quad \text{for all } R \in (0,\infty) \quad , \\ \dfrac{1}{1+|z|^2} \displaystyle\iint_{|z-\xi| \leq R} B(\xi,\omega) \, d\xi \, d\omega \to 0 \quad \text{as } |z| \to +\infty . \end{cases} \tag{8}$$

One famous example of such kernel B is the so-called hard-sphere model where $B(z,\omega) = |(z,\omega)|$. The assumption (8), even if it seems reasonable enough, is in fact a serious limitation in view of the violent singularities present in "realistic" collision kernels. And it corresponds to the so-called *angular cut-off* approximation which consists in neglecting grazing collisions (see H. Grad [**23**], C. Cercignani [**7**]...). More details on Boltzmann model can be found in [**7**],[**8**], C. Truesdell and R. Muncaster [**36**]...

The Boltzmann equation ((B) in short) corresponds to the general equation (1) with $F = 0$ and C given by (7). We can of course combine this model with the preceding ones, forming thus the Vlasov-Boltzmann model ((VB) in short) composed of (1),(2),(7), the Vlasov-Maxwell-Boltzmann model ((VMB) in short) composed of (1),(3),(4),(5) and (7)...

Another classical model for C was introduced by Landau (see [**26**], S. Chapman and T.G. Cowling [**8**]) and physically corresponds to emphasizing the role of grazing collisions (see L. Desvillettes [**10**], P. Degond and B. Lucquin-Desreux [**9**]...) :

$$C = Q(f,f) = \sum_{i,j=1}^{N} \frac{\partial}{\partial \xi_i} \left\{ \int_{\mathbb{R}^N} d\xi_* \, a_{ij}(\xi - \xi_*) \left[f(\xi_*) \frac{\partial f}{\partial \xi_j} - f(\xi) \frac{\partial f_*}{\partial \xi_{*,j}} \right] \right\} \tag{9}$$

where we always assume that $(a_{ij}) = (a_{ji})$ is even in z and satisfies at least

$$(10) \qquad \begin{cases} \sum_{i,j=1}^{N} a_{ij}(z) z_i z_j = 0 & \text{a.e. } z \in \mathbf{R}^N , \\ \sum_{i,j=1}^{N} a_{ij}(z) \eta_i \eta_j > 0 & \text{if } \eta \cdot z = 0 , \ \eta \neq 0 , \text{ a.e. } z \in \mathbf{R}^N . \end{cases}$$

A typical example when $N = 3$ is given by : $a_{ij}(z) = \frac{\delta_{ij}}{|z|} - \frac{z_i z_j}{|z|^3}$. The Landau model ((L) in short) corresponds to equation (1) with $F = 0$ and C given by (9).

Of course, all these systems have to be complemented with initial conditions

$$(11) \qquad f|_{t=0} = f_0 \quad \text{on} \quad \mathbf{R}^N_x \times \mathbf{R}^N_\xi .$$

And in the case when we use (3)-(4) ((VM) or (VMB) models), we have to prescribe (E, B) at time 0 i.e.

$$(12) \qquad E|_{t=0} = E_0 \quad , \quad B|_{t=0} = B_0 \quad \text{on} \quad \mathbf{R}^3$$

with the usual compatibility conditions

$$(13) \qquad \operatorname{div} B_0 = 0 \quad , \quad \operatorname{div} E_0 = \rho_0 = \int_{\mathbf{R}^3} f_0(x, \xi)\, d\xi .$$

Of course, f_0 is nonnegative and formally at least, all these systems preserve the nonnegativity of the initial condition i.e. $f \geq 0$ on $\mathbf{R}^N_x \times \mathbf{R}^N_\xi \times [0, \infty)$.

We next want to present the classical *a priori estimates* that are available for these models. And in order to simplify the presentation, we shall restrict our attention in *all* that follows to $N = 3$ even if all our results can be either extended or adapted to arbitrary dimensions. We only want to avoid technical (and irrelevant) digressions.

We begin with a priori estimates that are valid for all the models described above. Due to invariances (and symmetries), we find at least formally

$$(14) \qquad \frac{d}{dt} \iint_{\mathbf{R}^6} f \, dx \, d\xi = 0$$

and since $f \geq 0$, this provides a trivial L^1 bound ($L^\infty_t(L^1_{x,\xi})$ precisely). The conservation of total energy takes the following form : solutions satisfy again formally

$$(15) \qquad \frac{d}{dt} \left\{ \iint_{\mathbf{R}^6} f|\xi|^2 \, dx \, d\xi + \iint_{\mathbf{R}^6} \rho(x) V_0(x-y) \rho(y) \, dx \, dy \right\} = 0$$

if F is given by (2) (if $F \equiv 0$, we simply take $V_0 \equiv 0$) and

$$(16) \qquad \frac{d}{dt} \left\{ \iint_{\mathbf{R}^6} f|\xi|^2 \, dx \, d\xi + \int_{\mathbf{R}^3} |E|^2 + |B|^2 \, dx \right\} = 0$$

if F is given by (3). In the latter case, this yields bounds on f in $L^\infty_t(L^1_{x,\xi}(|\xi|^2))$ and on E, B in $L^\infty_t(L^2_x)$. In the Vlasov case (with, possibly, collision kernels...) we have to assume that V_0 is bounded from below in order to deduce that f is bounded in $L^\infty_t(L^1_{x,\xi}(|\xi|^2))$ and that $\rho(x)V_0^+(x-y)\rho(y)$ is bounded in $L^\infty_t(L^1_{x,y})$. The next bound we want to discuss comes from the increase of *entropy* described mathematically by the following identities

$$(17)\quad \frac{d}{dt}\iint_{\mathbb{R}^6} f\log f\,dx\,d\xi + \frac{1}{4}\int_{\mathbb{R}^3} dx \iint_{\mathbb{R}^6} d\xi\,d\xi_* \int_{S^2} d\omega\, B(f'f'_* - ff_*)\log\frac{f'f'_*}{ff_*} = 0$$

in the case of (7), and

$$(18)\quad \begin{cases} \dfrac{d}{dt}\displaystyle\iint_{\mathbb{R}^6} f\log f\,dx\,d\xi + \frac{1}{2}\int_{\mathbb{R}^3} dx \iint_{\mathbb{R}^6} d\xi\,d\xi_*\, ff_*\cdot \\ \qquad \displaystyle\sum_{i,j=1}^{3} a_{ij}\left(\frac{\partial}{\partial\xi_i}\log f - \frac{\partial}{\partial\xi_{*,i}}\log f_*\right)\left(\frac{\partial}{\partial\xi_j}\log f - \frac{\partial}{\partial\xi_{*,j}}\log f_*\right) = 0 \end{cases}$$

in the case of (9). Notice that the two terms that describe the dissipation of $(-)$ entropy are nonnegative. As such (17) and (18) do not really imply bounds on f or on $f\log f$ because of the negativity of log near 0 and we need some control, say, of f for large x.

This control at large x comes from another formal identity

$$(19)\quad \frac{d^2}{dt^2}\iint_{\mathbb{R}^6} f|x|^2\,dx\,d\xi = 2\iint_{\mathbb{R}^6}|\xi|^2 f\,dx\,d\xi - \iint_{\mathbb{R}^6}(x-y)\cdot\nabla V_0(x-y)\,\rho(x)\,\rho(y)\,dx\,dy$$

if F is given by (2). And if $V_0 \equiv 0$ i.e. we are considering either the (B) model or the (L) model, we can also rewrite (19) in order to obtain

$$(20)\quad \frac{d^2}{dt^2}\iint_{\mathbb{R}^6} f|x-\xi t|^2\,dx\,d\xi = 0\,.$$

These identities yield $L^\infty_t(L^1_{x,\xi}(|x|^2))$ bounds that can be used to bound $f|\log f|$ by the following observation due to L. Arkeryd [2]

$$\begin{aligned} \iint_{\mathbb{R}^6} f|\log f|\,dx\,d\xi &\le \iint_{\mathbb{R}^6} f\log f\,dx\,d\xi + 2\iint_{\mathbb{R}^6} f|\log f|\,1_{(f\le 1)}\,dx\,d\xi \\ &\le \iint_{\mathbb{R}^6} f\log f\,dx\,d\xi + 2\iint_{\mathbb{R}^6} f|\log f|\,1_{(f\le m)} + f|\xi|^2 + f|x-x_0|^2\,dx\,d\xi \end{aligned}$$

where $m = \exp\{-(|\xi|^2 + |x-x_0|^2)\}$ and $x_0 = x_0(\xi)$ is arbitrary in $\mathbb{R}^3$. In fact, we choose $x_0 = 0$ in the (VB) (or (VL)) case and $x_0 = t\xi$ in the (B) or (L) cases.

In order to summarize what we have achieved with all those identities, let us conclude this presentation by recalling the estimates shown : let $f_0 \geq 0$ satisfy for some $R \in [0, \infty)$

$$(21) \quad \iint_{\mathbb{R}^6} f_0 (1 + |x|^2 + |\xi|^2 + |\log f_0|) \, dx \, d\xi + \iint_{\mathbb{R}^6} \rho_0(x) \, V_0^+(x-y) \rho_0(y) \, dx \, dy \leq R$$

with the convention that $V_0 \equiv 0$ if we study the (B) or (L) models. Then, formally at least, if f is a solution of (B) or (L) satisfying (11), we find for some $C = C(R) \geq 0$

$$(22) \qquad \iint_{\mathbb{R}^6} f(t) \, (1 + |x - \xi t|^2 + |\xi|^2 + |\log f(t)|) \, dx \, d\xi \leq C$$

and

$$(23) \qquad \int_0^\infty dt \int_{\mathbb{R}^3} dx \iint_{\mathbb{R}^6} d\xi \, d\xi_* \, D \leq C$$

where $D = \int_{S^2} B(f'f'_* - ff_*) \log \frac{f'f'_*}{ff_*} \, d\omega$ in the (B) case, $D = ff_* \sum_{i,j=1}^3 a_{ij} \left(\frac{\partial}{\partial \xi_i} \log f - \frac{\partial}{\partial \xi_{*,i}} \log f_*\right) \left(\frac{\partial}{\partial \xi_j} \log f - \frac{\partial}{\partial \xi_{*,j}} \log f_*\right)$ in the (L) case.

Similarly, we find in the (VB) (or (VL)) case, provided V_0 satisfies for some $C \geq 0$

$$(24) \qquad V_0 \geq -C \quad , \quad -x \cdot \nabla V_0 \leq C(V_0^+ + 1) \qquad \text{in } \mathbb{R}^3 ,$$

that solutions satisfy for some $C = C(R) \geq 0$

$$(22') \qquad \begin{cases} \displaystyle\iint_{\mathbb{R}^6} f(t) \, (1 + |x|^2 + |\xi|^2 + |\log f(t)|) \, dx \, d\xi \\ \displaystyle\qquad + \iint_{\mathbb{R}^6} \rho(x,t) \, |V_0|(x-y) \, \rho(y,t) \, dx \, dy \leq C(1+t)^2 \end{cases}$$

and for all $T \in [0, \infty)$

$$(23') \qquad \int_0^T dt \int_{\mathbb{R}^3} dx \iint_{\mathbb{R}^6} d\xi \, d\xi_* \, D \leq C(1+T)^2 .$$

Remark II.1.
Notice that we excluded the case of the (VMB) model where identities like (19) are not clear. Estimates similar to (22'),(23') are possible nevertheless (assuming a bit more on B) but we de not want to enter those technicalities here. □

If no collision terms are present i.e. if we are considering the "pure" (V) or (VM) cases, additional estimates are available. Indeed, we observe that if $C = 0$ and f solves (1) then, for any $\beta \in C([0, \infty); \mathbb{R})$, $\beta(f)$ solves (1). This observation merely reflects the fact that if f is constant along particle paths then $\beta(f)$ is also constant along particle paths. Then, using (6), we deduce the following formal identity that corresponds to the famous Liouville invariance (of the Lebesgue measure)

$$\frac{d}{dt}\iint_{\mathbf{R}^6}\beta(f)\,dx\,d\xi = 0\,. \tag{25}$$

This immediately yields $L_t^\infty(L^p_{x,\xi})$ bounds for all $1 \leq p \leq \infty$.

In turn, these bounds can be used in order to bound the total energy and each part of it in cases when V_0 is *not* bounded from below. Indeed, one recalls first rather standard interpolation arguments that yield the

Lemma II.1.

Let g satisfy : $g \in L^p(\mathbf{R}^6_{x,\xi})$ for some $p \in [1,\infty]$, $g|\xi|^m \in L^1(\mathbf{R}^6_{x,\xi})$ for some $m > 0$, let $r \in (0,m)$. Then, $\rho = \int_{\mathbf{R}^3} |g|\,|\xi|^r\,d\xi \in L^q(\mathbf{R}^3_x)$ where $\frac{1}{q} = \theta + \frac{1-\theta}{p}$, $\theta = \left(\frac{3}{p'}+r\right)\left(\frac{3}{p'}+m\right)^{-1}$ and we have for some $C \geq 0$ independent of g

$$\|\rho\|_{L^q_x} \leq C\,\|g\|^{1-\theta}_{L^p_{x,\xi}}\,\|g|\xi|^m\|^{\theta}_{L^1_{x,\xi}}\,. \tag{26}$$

Proof.

This follows directly from interpolation results presented for instance in Berg-Lofström [4]. A direct (and standard) proof consists in writing for all $x \in \mathbf{R}^3$

$$\rho(x) = \int_{|\xi|\leq R}|g|\,|\xi|^r\,d\xi + \int_{|\xi|>R}|g|\,|\xi|^r\,d\xi$$

where $B \geq 0$ will be determined later on. Then, we estimate ρ as follows

$$\rho(x) \leq C\left(\int_{\mathbf{R}^3_\xi}|g|^p\,d\xi\right)^{1/p} R^{\frac{3}{p'}+r} + \frac{1}{R^{m-r}}\left(\int_{\mathbf{R}^3_\xi}|g|\,|\xi|^m\,d\xi\right). \tag{27}$$

But, $g_1 = C\left(\int_{\mathbf{R}^3_\xi}|g|^p\,d\xi\right)^{1/p} \in L^p(\mathbf{R}^3_x)$, $g_2 = \int_{\mathbf{R}^3_\xi}|g|\,|\xi|^m\,d\xi \in L^1(\mathbf{R}^3_x)$ and minimizing (at each $x \in \mathbf{R}^3$) the right-hand side of (27) with respect to R we deduce

$$\rho \leq g_1^{1-\theta}\,g_2^{\theta} \quad \text{on } \mathbf{R}^3\,.$$

And the lemma is proven. □

Next, we may use this lemma to bound the negative part of the total energy in (V) models as follows. We assume that the interaction potential satisfies

$$V_0^- \in L^\infty + L^{\alpha,\infty}\quad,\quad \text{with } \alpha > \frac{3}{2}\quad (L^{\alpha,\infty} = L^\infty \text{ if } \alpha = \infty) \tag{28}$$

where $L^{\alpha,\infty}$ denotes the weak L^α space (or Marcinkiewicz space).

Then, using (25), we may assume $L^p_{x,\xi}$ bounds uniform in t for $1 \le p \le \overline{p}(\alpha)$ where $\overline{p} = \frac{4\alpha-3}{4\alpha-5}$. Notice that $\overline{p} < 3$ since $\alpha > \frac{3}{2}$. Then, applying Lemma II.1 with $m = 2$, $r = 0$, $p = \overline{p}$, we obtain L^q_x bounds on ρ uniform in t for $1 \le q \le \overline{q} = \frac{2\alpha}{2\alpha-1}$ and we have

$$\|\rho\|_{L^{\overline{q}}} \le C\left(\iint_{\mathbb{R}^6_{x,\xi}} f|\xi|^2\,dx\,d\xi\right)^{\theta}$$

where $\theta = \frac{3}{4\alpha}$. Notice that $\theta < \frac{1}{2}$ since $\alpha > \frac{3}{2}$.

Next, using (28), we deduce

$$\iint_{\mathbb{R}^3\times\mathbb{R}^3} \rho(x)\,V_0^-(x-y)\,\rho_0(y)\,dx\,dy \le C + C\left(\iint_{\mathbb{R}^6} f|\xi|^2\,dx\,d\xi\right)^{2\theta}.$$

This equality combined with (15) yields the desired bounds.

In conclusion, we have shown the following facts for the (V) models. Let $f_0 \ge 0$ satisfy for some $R \ge 0$ and for some $p \in [\overline{p}, +\infty]$

$$(29)\quad \iint_{\mathbb{R}^6} f_0(1+|\xi|^2)\,dx\,d\xi + \|f_0\|_{L^p_{x,\xi}} + \iint_{\mathbb{R}^3\times\mathbb{R}^3} \rho_0(x)\,V_0^+(x-y)\,\rho_0(y)\,dx\,dy \le R\,.$$

Then, if V_0 satisfies (28), f solution of (V) satisfying (11) admits, at least formally, the following bounds

$$(30)\quad \iint_{\mathbb{R}^6} f(1+|\xi|^2)\,dx\,d\xi + \|f\|_{L^p_{x,\xi}} + \iint_{\mathbb{R}^3\times\mathbb{R}^3} \rho(x)\,|V_0(x-y)|\,\rho(y)\,dx\,dy \le C(R)$$

for all $t \in [0,\infty)$, and for some constant $C(R) \ge 0$.

Remark II.2.
Let $V_0(x) = -\frac{1}{|x|}$ ((VP$_-$) model, then (28) holds with $\alpha = 3$ since $V_0 \in L^{3,\infty}(\mathbb{R}^3)$. In that case, $\overline{p} = \frac{9}{7}$.

Remark II.3.
The condition (28) excludes potentials V_0 like : $V_0(x) = -\frac{1}{|x|^\beta}$ for $\beta \in [2,3)$. For those potentials, one can show using (29) blow-up results that prove that smooth solutions do not exist for all $t \ge 0$ in general and that the above estimates are false. □

We learn from this analysis that writing the equation satisfied by $\beta(f)$ can provide some extra bounds and this is one of the key facts that we shall use on the Boltzmann equation. Indeed, we conclude this section by going back to the Boltzmann model ((B) or even (VB), (VMB) models). Then, if we write the equation satisfied by $\beta(f) = \log(1+f)$ (for instance) we find using (6)

$$(31)\qquad \frac{\partial}{\partial t}\beta(f) + \mathrm{div}_x(\xi\beta(f)) + \mathrm{div}_\xi(F\beta(f)) = \beta'(f)\,Q(f,f)\,.$$

And $\beta'(f)\,Q(f,f) = \frac{1}{1+f}[Q_+(f,f) - Q_-(f,f)] = \frac{1}{1+f}\,Q_+(f,f) - \frac{f}{1+f}\,L(f)$ where

$$\text{(32)}\qquad \begin{cases} Q_+(f,f) \;=\; \displaystyle\iint_{\mathbb{R}^3\times S^2} B(\xi-\xi_*,\omega)\, f' f'_*\, d\xi_*\, d\omega\;, \\ Q_-(f,f) \;=\; f\,L(f)\quad,\quad L(f) \;=\; \displaystyle\iint_{\mathbb{R}^3\times S^2} B\, f_*\, d\xi_*\, d\omega\;. \end{cases}$$

Next, we observe that because of (8), (22) or (22'), we have respectively in the (B) case or (VB), (VMB) cases

$$\text{(33)}\qquad \frac{1}{1+f}\,Q_-(f,f) \;\in\; L^\infty(0,\infty\,;\,L^1(\mathbb{R}^3_x\times B_\xi))\quad,\quad \text{for any ball}\quad B\subset\mathbb{R}^3$$

or

(33')
$$\frac{1}{1+f}\,Q_-(f,f) \;\in\; L^\infty(0,T\,;\,L^1(\mathbb{R}^3_x\times B_\xi))\,,\quad \text{for any } T\in(0,\infty)\,,\ \text{ball}\quad B\subset\mathbb{R}^3\,.$$

We then wish to derive similar bounds on $Q_+(f,f)$. In order to do so, we pick $R\in(0,\infty)$, choose $\varphi\in C_0^\infty(\mathbb{R}^3)$ such that $\varphi\geq 0$, $\varphi\equiv 1$ on $\{|\xi|<R\}$ and we multiply (31) by $\varphi(\xi)$ and integrate over $[0,T]\times\mathbb{R}^3_x\times\mathbb{R}^3_\xi$. We then find using (33),(33'),(22),(22')

$$\int_0^T dt\int_{\mathbb{R}^3} dx\int_{(|\xi|<R)}\frac{Q_+(f,f)}{1+f}\,d\xi\;\leq\; C(R,T)+\int_0^T dt\int_{\mathbb{R}^6} dx\,d\xi\,|F\cdot\nabla\varphi(\xi)|\,\beta(f)\,.$$

We observe that $\beta(f)$ is bounded (uniformly in t) in $L^q_{x,\xi}$ for all $1\leq q<\infty$. In the (B) case $F\equiv 0$, in the (VMB) case $\frac{F}{1+|\xi|}$ is bounded in $L^\infty_t(L^2_x)$ because of (16), while in the (VB) case if V_0 satisfies

$$\text{(34)}\qquad \nabla V_0\;\in\; L^\infty + L^p\qquad \text{for some}\quad p>1\,,$$

then $F=\nabla V_0 * \rho$ is bounded in $L^\infty_t(L^\infty+L^p)$. In all cases, we deduce

$$\text{(35)}\qquad \frac{1}{1+f}\,Q_+(f,f)\;\in\; L^1(0,T;L^1(\mathbb{R}^3_x\times B_\xi))\qquad \text{for any } T\in(0,\infty)\,,\ \text{ball } B\subset\mathbb{R}^3\,.$$

In fact, (33), (33'), (35) are a priori bounds that depends only on L^1 bounds on f (and $f|\xi|^2$), bounds that we already obtained above. Let us also observe that choosing β differently (bounded), (34) may be replaced by

$$\text{(34')}\qquad \nabla V_0\;\in\; L^\infty + L^1\,.$$

III. Existence and convergence results.

We begin by stating existence results of global weak solutions for the (VB) model, then the (V) model and finally the (VM) model. Next, in the second part of this section we explain how these existence results are deduced from stability results.

First, we define *renormalized solutions* of (VB) (and of (V) !) : f is said to be a renormalized solution of (VB) if $f \in C([0,\infty); L^1(\mathbb{R}^6_{x,\xi}))$; $f(|x|^2 + |\xi|^2 + |\log f|) \in L^\infty(0,T; L^1(\mathbb{R}^6_{x,\xi}))$ (for all $T \in (0,\infty)$) ; $\frac{Q^-(f,f)}{1+f} \in L^\infty(0,T; L^1(\mathbb{R}^3_x \times B_\xi))$, $\frac{Q^+(f,f)}{1+f} \in L^1(0,T; L^1(\mathbb{R}^3_x \times B_\xi))$ for all $T \in (0,\infty)$), balls $B \subset \mathbb{R}^3$; $\beta(f) = \log(1+f)$ satisfies (31) in the sense of distributions and we have

$$\text{(E)} \qquad \iint_{\mathbb{R}^6} f \log f \, dx\, d\xi(t) + \int_0^t ds \int_{\mathbb{R}^3} dx \iint_{\mathbb{R}^6} d\xi\, d\xi_*\, D \;\le\; \iint_{\mathbb{R}^6} f \log f\, dx\, d\xi(0)$$

where $D = \int_{S^2} B(f'f'_* - ff_*) \log \frac{f'f'_*}{ff_*}\, d\omega$.

We recall that we assume (8), and that V_0 satisfies (24). We shall also assume

$$\text{(36)} \qquad \nabla V_0 \in L^\infty + L^{3/2,\infty} ,$$

$$\text{(37)} \qquad D^2 V_0 \in L^\infty + \mathcal{M} + K ,$$

and a technical condition on V_0 (that might not be necessary for our analysis...)
(38)

$$\begin{cases} \text{if } \varphi \ge 0 , \quad \int_{\mathbb{R}^3} \varphi(1+|x|^2+|\log \varphi|)\, dx + \iint_{\mathbb{R}^3\times\mathbb{R}^3} \varphi(x)\, V_0^+(x-y)\, \varphi(y)\, dx\, dy < \infty \\ \text{then} \quad \nabla V_0 * \varphi \in L^{\frac{5}{3}} + L^\infty(\mathbb{R}^3) . \end{cases}$$

Here and below $\mathcal{M}$ is the space of bounded measures and K is a space of tempered distributions we define as follows : $T \in K$ if the convolution operator $(T*)$ is bounded on $L^p(\mathbb{R}^3)$ for all $p \in (1,\infty)$ and on the Hardy space $H_1(\mathbb{R}^3)$. Conditions on T that ensure these properties are rather technical and we skip them here... A typical and important example is given by the Coulomb potential $V_0(x) = \frac{1}{|x|}$. In that case, (36) holds since $\nabla V_0 \in L^{3/2,\infty}$ and $(\frac{\partial^2}{\partial x_i\, \partial x_j} V_0 *)$ is nothing but $R_i R_j$ ($\forall\, i,j$) where $(R_i)_i$ and the Riesz transforms. Therefore, the Coulomb potential does satisfy (36)-(37). Another example is the Yukawa potential $V_0(x) = \frac{e^{-\mu|x|}}{|x|}$ for some $\mu > 0$... Notice also that (36) is essentially implied by (37) in view of Sobolev embeddings.

Let us also observe that the technical condition (38) holds in those two examples since $V_0 = V_0^+$ and $\iint_{\mathbb{R}^3\times\mathbb{R}^3} \varphi(x)\, V_0(x-y)\, \varphi(y)\, dx\, dy \ge 4\pi \,\|\nabla V_0 * \varphi\|^2_{L^2(\mathbb{R}^3)}$ in both cases (equality holds if $V_0 = \frac{1}{|x|}$). Then, (38) holds since $\nabla V_0 * \varphi \in L^{3/2,\infty}(\mathbb{R}^3)$ if $\varphi \in L^1(\mathbb{R}^3)$.

Finally, the initial condition f_0 is a nonnegative function that is assumed to satisfy

$$(39)\quad \iint_{\mathbb{R}^6} f_0\,(1+|x|^2+|\xi|^2+|\log f_0|)\,dx\,d\xi+\iint_{\mathbb{R}^3\times\mathbb{R}^3}\varphi_0(x)\,V_0^+(x-y)\,\rho_0(y)\,dx\,dy\;<\;\infty\,.$$

Theorem III.1.

With the above notations and conditions, there exists a renormalized solution of (VB) satisfying (11).

Remark III.1.

In the case when $V_0 \equiv 0$ (the (B) model), this result is taken from R.J. DiPerna and P.L. Lions [**11**],[**12**].

Remark III.2.

Additional regularity properties of solutions and their uniqueness are outstanding open problems.

Remark III.3.

A similar result is not known for the (VMB) model even if some of the compactness properties stated in the stability results presented below remain true.

We next consider the (V) model and we first define (even if we already did above!) renormalized solutions of (V) : f is said to be a renormalized solution of (V) if $f \in C([0,\infty);L^1(\mathbb{R}^6_{x,\xi}))$, $f|\xi|^2 \in L^\infty(0,\infty;L^1(\mathbb{R}^6_{x,\xi}))$, $(\rho * |V_0|)\rho \in L^\infty(0,\infty;L^1(\mathbb{R}^3_x))$ and for each $\beta \in C([0,\infty);\mathbb{R})$ bounded on $[0,\infty)$, $\beta(f)$ satisfies (31) in the sense of distributions (with $Q \equiv 0$). We shall assume that the initial condition f_0 is nonnegative and satisfies

$$(40)\quad \iint_{\mathbb{R}^6} f_0(1+|\xi|^2)\,dx\,d\xi+\iint_{\mathbb{R}^3\times\mathbb{R}^3}\rho_0(x)\,V_0^+(x-y)\,\rho_0(y)\,dx\,dy+\|f_0\|_{L^p_{x,\xi}}\;<\;\infty$$

for some $p \in [\overline{p},+\infty]$ and α is the exponent introduced in (28). We agree that if $\alpha = +\infty$ (so that $\overline{p} = 1$) and if $p = 1$, we replace $\|f_0\|_{L^1_{x,\xi}}$ by $\iint_{\mathbb{R}^6_{x,\xi}} f_0 \log^+ f_0\,dx\,d\xi$. The last assumption we use is a technical condition on V_0 (that probably can be relaxed) which extends the condition (38)

$$(41)\quad \begin{array}{l}\text{if}\ \ \varphi \geq 0\ \ ,\ \ \varphi \in L^1 \cap L^q(\mathbb{R}^3)\ \ ,\ \ (\varphi * V_0^+)\varphi \in L^1(\mathbb{R}^3)\ \ ,\\ \qquad\qquad \text{then}\ \ \nabla V_0 * \varphi \in L^{5/3}+L^\infty(\mathbb{R}^3)\end{array}$$

where q is given by (see Lemma II.1) : $\frac{1}{q} = \theta+\frac{1-\theta}{p}$, $\theta = \frac{3}{p'}\left(\frac{3}{p'}+2\right)^{-1}$ and again we agree that if $p = 1$ (so $q = 1$) then we replace $\varphi \in L^1 \cap L^q(\mathbb{R}^3)$ by $\int_{\mathbb{R}^3}\varphi(1+\log^+\varphi)\,dx < \infty$. Observe that if we choose $p = +\infty$ in (39), then $q = \frac{5}{3}$ so that (41) holds as soon as $\nabla V_0 \in L^1 + L^\infty(\mathbb{R}^3)$!

Theorem III.2.

We assume (28),(36),(37),(40) and (41).

(i) Then, there exists a renormalized solution f of (V) satisfying (11). In addition, $\{f(t) \mid t \in [0,\infty)\}$ is equimeasurable and $f \in C([0,\infty); L^q_{x,\xi})$ if $f_0 \in L^q_{x,\xi}$ and $1 \le q < \infty$ while f is bounded on $\mathbf{R}^6_{x,\xi} \times [0,\infty)$ if f_0 is bounded.

(ii) If f^0 satisfies in addition : $f^0 \in L^\infty(\mathbf{R}^6_{x,\xi})$ and we have for some $m_0 > 3$

$$\iint_{\mathbf{R}^6} |\xi|^m f_0 \, dx \, d\xi < \infty \qquad \text{for} \quad 0 \le m < m_0 , \tag{42}$$

then f satisfies for all $T \in (0,\infty)$

$$\sup_{t\in[0,T]} \iint_{\mathbf{R}^6} |\xi|^m f(t) \, dx \, d\xi < \infty \qquad \text{for} \quad 0 \le m < m_0 , \tag{43}$$

and

$$\rho \in C([0,\infty); L^q(\mathbf{R}^3)) \qquad \text{for} \quad 1 \le q < \frac{3+m_0}{3} . \tag{44}$$

Finally, if $m_0 > 6$ and f_0 satisfies for all $R, T \in (0,\infty)$

$$\operatorname{ess\,sup} \left\{ f_0(y+\xi t, \eta) \; / \; |y-x| \le Rt^2 \, , \, |\xi-\eta| \le Rt \right\} \in L^\infty((0,T)\times \mathbf{R}^3_x; L^1(\mathbf{R}^3_\xi)) \tag{45}$$

then $\rho \in L^\infty(\mathbf{R}^3_x \times (0,T))$ for all $T \in (0,\infty)$.

Remark III.4.

Part (i) is an adaptation of R.J. DiPerna and P.L. Lions [**14**] while part (ii) is an adaptation of B. Perthame and P.L. Lions [**30**]. These adaptations simply consist in rewriting the proofs for general potentials V_0 instead of $V_0 = \pm\frac{1}{|x|}$. Of course, the bounds given in (ii) imply regularity and uniqueness results (see [**39**]) that are somewhat more general than those independently obtained by K. Pfaffelmoser [**34**], R. Glassey[**20**], J. Schaeffer [**35**].

The final existence result that we want to mention is taken from R.J. DiPerna and P.L. Lions [**15**] and concern the (VM) model.

Theorem III.3.

Let $f_0 \ge 0$ belong to $L^1(\mathbf{R}^6, |\xi|^2 \, dx \, d\xi) \cap L^2(\mathbf{R}^6_{x,\xi})$, let $E_0, B_0 \in L^2(\mathbf{R}^3_x)$. We assume that (13) holds (in the sense of distributions). Then, there exists $f \in L^\infty(0,\infty; L^1(\mathbf{R}^6, |\xi|^2 \, dx \, d\xi)) \cap L^2(\mathbf{R}^6_{x,\xi}))$, $E, B \in L^\infty(0,\infty; L^2(\mathbf{R}^3_x))$ solution of (VM) which satisfy (11) and (12), and such that $f \in L^\infty(0,\infty; L^p(\mathbf{R}^6_{x,\xi}))$ if $f_0 \in L^p(\mathbf{R}^6_{x,\xi})$ for any $p \in [1,\infty]$.

Remark III.7.

The equations composing (VM) hold in the sense of distributions and the term $F \cdot \nabla_\xi f$ is interpreted as $\operatorname{div}_\xi(Ff)$ while $Ff \in L^\infty(0,\infty; L^1_{\text{loc}})$ in view of the bounds satisfied by f, E, B. Regularity and uniqueness are important open problems and the existence of a renormalized solution is not known. The solution we build satisfies $f(t) \prec f_0$ for all $t \geq 0$ where $\prec$ denotes the Hardy-Littlewood domination relationship :

$$\int_{\mathbb{R}^6_{x,\xi}} F(f(t))\,dx\,d\xi \leq \int_{\mathbb{R}^6_{x,\xi}} F(f_0)\,dx\,d\xi$$

for all $t \geq 0$ and for all F continuous, nonnegative, convex on $[0,\infty)$.

In all the global existence results (of somewhat weak solutions) stated above, one of the crucial facts in the proofs is the ability to pass to the limit in the equations when one considers bounded sequences of solutions. By bounded solutions, we mean the bounds shown in the preceding section II. These bounds only yield weak compactness which is not, a priori, sufficient to pass to the limit in the nonlinear terms. As it is often the case for nonlinear PDE's, we need some additional compactness informations in order to make this passage to the limit possible. This extra-compactness and the stability that we can obtain will be the subject of the rest of this section.

We begin with the (VB) and (B) cases

Theorem III.4.

Under the assumptions of Theorem III.1, let $(f^n)_n$ be a sequence of renormalized solutions of (VB) satisfying for all $T \in (0,\infty)$

$$(46)\qquad \begin{cases} \sup_{n,t\in[0,T]} \Big[\iint_{\mathbb{R}^6} f^n(t)(1+|x|^2+|\xi|^2+|\log f^n(t)|)\,dx\,d\xi \\ \qquad + \iint_{\mathbb{R}^3\times\mathbb{R}^3} \rho^n(x,t)\,V_0^+(x-y)\,\rho^n(y,t)\,dx\,dy < \infty \Big]. \end{cases}$$

Without loss of generality, we may assume that f^n converges weakly in $L^1(\mathbb{R}^6_{x,\xi}\times(0,T))$ (for all $T \in (0,\infty)$) to some $f \geq 0$. Then, f is a renormalized solution of (VB) (satisfying (E) with $\iint_{\mathbb{R}^6} f(0)\log f(0)\,dx\,d\xi$ replaced by $\underline{\lim}_n \iint_{\mathbb{R}^6} f^n(0)\log f^n(0)\,dx\,d\xi$). And we have for all $\psi \in L^\infty(\mathbb{R}^3_\xi)$, $T \in (0,\infty)$, $p \in [1,\infty)$

$$(47)\qquad \int_{\mathbb{R}^3} f^n(x,\xi,t)\,\psi(\xi)\,d\xi \underset{n}{\rightarrow} \int_{\mathbb{R}^3} f(x,\xi,t)\,\psi(\xi)\,d\xi \quad \text{in} \quad L^p(0,T;L^1(\mathbb{R}^3_x)).$$

Remark III.8.

In the result above, one can show that $f(0)$ is the weak limit in $L^1(\mathbb{R}^6_{x,\xi})$ of $f^n(0)$. Notice also that (47) is in general false with a convergence in $C([0,T];L^1(\mathbb{R}^3_x))$ since the behaviour of $f^n(0)$ can be almost arbitrary (only bounds are assumed). The physical meaning of (47) is clear : averages in ξ represent macroscopic quantities and (47) means the oscillations (or fluctuations) that can be present in the full densities are averaged out in macroscopic quantities ! □

We now turn to the (V) case and state a result adapted from R.J. DiPerna and P.L. Lions [**14**].

Theorem III.5.

We assume (28),(36),(37) and (41). Let $(f^n)_n$ be a sequence of renormalized solutions of (V) satisfying

$$(47)\quad \begin{cases} \sup_{t\geq 0,n} \iint_{\mathbb{R}^6} f^n(t)(1+|\xi|^2)\,dx\,d\xi \\ \qquad + \iint_{\mathbb{R}^3\times\mathbb{R}^3} \rho^n(x,t)\,V_0^+(x-y)\,\rho^n(y,t)\,dx\,dy + \|f^n(t)\|_{L^p_{x,\xi}} < \infty \end{cases}$$

with the same conditions on p as in (40). Without loss of generality, we may assume that f^n converges weakly in $L^1_{\text{loc}}(\mathbb{R}^6_{x,\xi}\times(0,\infty))$ to some $f\geq 0$.

(i) Then, f is an renormalized solution of (V) and we have for all $\psi\in L^\infty(\mathbb{R}^3_\xi)$, $R,T\in(0,\infty)$, $q\in[1,\infty)$

$$(47')\quad \int_{\mathbb{R}^3} f^n(x,\xi,t)\,\psi(\xi)\,d\xi \underset{n}{\rightarrow} \int_{\mathbb{R}^3} f(x,\xi,t)\,\psi(\xi)\,d\xi \quad \text{in } L^q(0,T;L^1(B_R)).$$

(ii) If $f^n(0)$ converges a.e. to f_0 on $\mathbb{R}^6_{x,\xi}$, then f^n converges a.e. to f on $\mathbb{R}^6_{x,\xi}\times[0,\infty)$ which satisfies (11).

(iii) If $f^n(0)$ converges to f_0 in $L^r(\mathbb{R}^6_{x,\xi})$ for some $1\leq r<\infty$ then f^n converges to f in $C([0,T];L^r(\mathbb{R}^6_{x,\xi}))$ (for all $T\in(0,\infty)$).

We then consider the (VM) model and we recall a result taken from R.J. DiPerna and P.L. Lions [15]

Theorem III.6.

Let $(f^n,E^n,B^n)_n$ be a sequence of weak solutions of (VM) satisfying for some $p\in(2,\infty)$

(48)

$$\sup_{t\geq 0,n} \iint_{\mathbb{R}^6} f^n(t)(1+|\xi|^2)\,dx\,d\xi + \|f^n(t)\|_{L^p(\mathbb{R}^6_{x,\xi})} + \|E^n(t)\|_{L^2(\mathbb{R}^3_x)} + \|B^n(t)\|_{L^2(\mathbb{R}^3_x)} < \infty .$$

Without loss of generality, we may assume that (f^n,E^n,B^n) converge weakly in L^2 to some (f,E,B). Then, (f,E,B) is a weak solution of (VM) and (47') holds.

We wish to mention a final compactness result valid for the (L) model (or for (VL) models...). This is the only result that we mention for this model to stress the main difference with the Boltzmann case. This result is taken from P.L. Lions [28] where renormalized solutions of (L) are considered and whose precise definitions will not be recalled here. We thus consider a sequence $(f^n)_n$ of renormalized solutions of (L) satisfying

$$(49)\quad \sup_{t\geq 0,n} \iint_{\mathbb{R}^6} f^n(t)\left(1+|x-\xi t|^2+|\xi|^2+|\log f^n(t)|\right)dx\,d\xi < \infty .$$

We shall assume that the matrix $(a_{ij}(z)) = (a_{ji}(z))$ satisfies (10) and

$$a_{ij} \in L^1 + L^\infty \quad (\forall\, i,j) \quad ; \quad \frac{\partial a_{ij}}{\partial z_j}, \frac{\partial^2 a_{ij}}{\partial z_i \partial z_j} \in \mathcal{M} + L^\infty . \tag{50}$$

Theorem III.7.
Under the above assumptions, $(f_n)_n$ is relatively compact in $L^p(0,T;L^1(\mathbf{R}^6_{x,\xi}))$ for all $1 \le p < \infty$, $T \in (0,\infty)$.

Remark III.9.
Notice that $a_{ij} = \frac{1}{|z|}\left(\delta_{ij} - \frac{z_i z_j}{|z|^2}\right)$ satisfies (50).

Remark III.10.
It is also shown in [28] that such a result is false for Boltzmann equations. And the proof in [28] can be adapted to (VL) or even (VML) models. □

Let us finally mention, without stating precise results, stability results involving weak solutions of (VM) (as obtained in Theorem III.3) and letting c go to $+\infty$. We, of course, expect to recover the (VP) model. This is something one can justify, by a straightforward adaptation of the proofs of the above results. And one deduces that if (f_0, E_0, B_0) satisfy (13) and $\int_{\mathbf{R}^6} f_0(1+|\xi|^2)\,dx\,d\xi + \|E_0\|_{L^2} + \|B_0\|_{L^2} + \|f_0\|_{L^p_{x,\xi}} < \infty$, for some $p \in [2,\infty)$ and if we denote by (f^c, E^c, B^c) a global weak solution of (VM) satisfying (11)-(12), then, extracting subsequences if necessary, f^c converges as c goes to $+\infty$ in $C([0,T];L^q(\mathbf{R}^6_{x,\xi}))$ for $1 \le q \le p$ ($\forall\, T \in (0,\infty)$) to a renormalized solution of (VP) satisfying (11). In addition, E^c, B^c converge weakly in $L^\infty(0,T;L^2_x) - w*$ ($\forall\, T \in (0,\infty)$) respectively to $-\Delta(\frac{1}{|x|} * \rho)$, 0. One can also recover smooth solutions, stronger convergences and allow for initial conditions that depend on c...

IV. Velocity averaging.

We have seen in the stability results of the preceding section that averages in ξ i.e. velocity averages of solutions of kinetic models are relatively compact. We want to explain these phenomena in this section by recalling a few results on velocity averages of linear transport equations.

We begin by explaining a general set-up : we consider functions g defined on $\mathbf{R}^N_x \times \mathbf{R}^k_v$ or $\mathbf{R}^N_x \times \mathbf{R}^k_v \times \mathbf{R}_t$ where $1 \le k, N$ satisfying

$$a(v) \cdot \nabla_x g = T \quad \text{in} \quad \mathcal{D}'(\mathbf{R}^N_x \times \mathbf{R}^k_v) \tag{51}$$

or

$$\frac{\partial g}{\partial t} + a(v) \cdot \nabla_x g = T \qquad \text{in} \quad \mathcal{D}'(\mathbf{R}_x^N \times \mathbf{R}_v^k \times \mathbf{R}_t)\,. \tag{52}$$

Here, a is a function from $\mathbf{R}^k$ into $\mathbf{R}^N$ and we assume, to simplify the presentation, that $a \in C^\infty$. Our first result is a slight improvement and extension of results shown in F. Golse, P.L. Lions, B. Perthame and R. Sentis **[21]** (previous remarks in that direction were obtained in F. Golse, B. Perthame and R. Sentis **[22]**, V. Agoshkov **[1]**).

Theorem IV.1.

We assume that $g \in L^p_{\mathrm{loc}}$ *solves (51) (resp. (52)) with* $T \in L^p_{\mathrm{loc}}$ and $1 < p < \infty$. *In addition, we assume that* a *satisfies for some* $\alpha \in (0,1]$

$$\begin{cases} \forall\, R \in (0,\infty)\,,\ \exists\, C > 0\,,\ \forall\, \delta \geq 0\,, \\ \text{meas}\left\{|v| \leq R\,,\ |a(v)\cdot\eta| \leq \delta\right\} \leq C\,\delta^\alpha \quad \text{for all}\ \eta \in S^{N-1} \end{cases} \tag{53}$$

(resp.

$$\begin{cases} \forall\, R \in (0,\infty)\,,\ \exists\, C > 0\,,\ \forall\, \delta \geq 0\,, \\ \text{meas}\left\{|v| \leq R\,,\ |\tau + a(v)\cdot\eta| \leq \delta\right\} \leq C\,\delta^\alpha \quad \text{for all}\ (\tau,\eta) \in S^N \quad). \end{cases} \tag{54}$$

Let $\psi \in L^\infty_{\mathrm{loc}}(\mathbf{R}^k)$ *with compact support. Then,* $\int_{\mathbf{R}^k} \psi(v) g\, dv \in H^{s,p}_{\mathrm{loc}}$ *where* $s = \frac{\alpha}{q}$ *and* $q = \max(p, p')$.

Remark IV.1.

The above result is known to be optimal when $p = 2$ and $a(v) = v \in \mathbf{R}^N$ (so $\alpha = 1$).

Example IV.1.

If we choose $k = N$, $v = \xi$, $a(v) = \xi$ then (53) and (54) hold with $\alpha = 1$.

Remark IV.2.

Let us explain briefly why it is natural to expect some improved regularity on $\int_{\mathbf{R}^k} \psi\, g\, dv$ in the case when $p = 2$. Indeed, one takes the Fourier transform in x of (51) for example and one deduces with obvious notations

$$(\eta \cdot a(v))\hat{g} = h \ \in\ L^2(\mathbf{R}^N \times \mathbf{R}^k)\,. \tag{55}$$

We are assuming here that $g, T \in L^2(\mathbf{R}^N \times \mathbf{R}^k)$, a global assumption that we can obtain from the local assumption by standard multiplications by cut-off functions.

Therefore, if $|\frac{\eta}{|\eta|} \cdot a(v)| \geq \delta$, we gain from (55), a decay of $\hat{g}$ of order $|\eta|^{-1}$. On the other hand, (53) means that the set of ("bad") v on which $|\frac{\eta}{|\eta|} \cdot a(v)| \leq \delta$, $v \in \operatorname{Supp} \psi$ is small for δ small and thus contributes in a controlled way to the average $\int_{\mathbb{R}^k} \psi(v)\hat{g}\,dv$. More precisely, we deduce from the above considerations

$$\begin{aligned}\left|\int_{\mathbb{R}^k} \psi(v)\hat{g}\,dv\right| &\le \left|\int_{\mathbb{R}^k} \psi(v)\,\frac{h}{\eta\cdot a(v)}\,1_{(|\eta\cdot a(v)|\ge\delta|\eta|)}\,dv\right| + C\,\delta^{\alpha/2}\left(\int_{\mathbb{R}^k}|\hat{g}|^2\,dv\right)^{1/2}\\ &\le \left(\int_{\mathbb{R}^k}|h|^2\,dv\right)^{1/2}\frac{1}{|\eta|}\left(\int_{|v|\le R}\left(\frac{\eta}{|\eta|}\cdot a(v)\right)^{-2} 1_{(|\eta\cdot a(v)|\ge\delta|\eta|)}\,dv\right)^{1/2} + h_2\,\delta^{\alpha/2}\\ &\le h_1\,\frac{1}{|\eta|}\,\delta^{\frac{\alpha}{2}-1} + h_2\,\delta^{\frac{\alpha}{2}}\ , \qquad \text{where } h_1,h_2\in L^2(\mathbb{R}^N_\eta)\,.\end{aligned}$$

Indeed the integral $\int_{|v|\le R}\left(\frac{\eta}{|\eta|}\cdot a(v)\right)^{-2} 1_{(|\eta\cdot a(v)|\ge\delta|\eta|)}\,dv$ can be written as

$$\int_\delta^\infty \frac{1}{t^2}\,d\mu(t) = \frac{1}{\delta^2}\,\mu(\delta) + 2\int_\delta^\infty \frac{\mu(t)}{t^3}\,dt \le C\,\delta^{\alpha-2}$$

where $\mu(t) = \sup_{\xi\in S^{N-1}}\left[\text{meas}\left\{|v|\le R,\, a(v)\cdot\xi|\le t\right\}\right] \le Ct^\alpha$ in view of (53).

The above bound yields for all $\eta\in\mathbb{R}^N$

$$\left|\int_{\mathbb{R}^k}\psi(v)\hat{g}\,dv\right| \le 2\,h_1^{\alpha/2}\,h_2^{1-\alpha/2}\,|\eta|^{-\alpha/2}$$

and the theorem is proved in the case $p = 2$. □

In fact, much more general regularity results are available and a complete theory exists now. We refer the reader to R.J. DiPerna, P.L. Lions and Y. Meyer [**18**]. Instead of presenting some of the results in [**18**], we prefer to state an adaptation of the results and methods of [**18**] which is only concerned with compactness.

Theorem IV.2.

We assume that g_n is bounded in L^p_{loc} and solves (51) (resp. (52)) with $T_n = (-\Delta_x+1)^{\tau/2}\cdot(-\Delta_v+1)^{m/2}\,G_n$ (resp. $(-\Delta_{x,t}+1)^{\tau/2}\cdot(-\Delta_v+1)^{m/2}\,G_n$) where G_n is a locally bounded measure uniformly in n, where $m\ge 0$, $\tau\in[0,1)$.

In addition, we assume that $\psi\in C_0^\infty(\mathbb{R}^k_v)$ and that a satisfies

(56) $$\text{meas}\left\{v\in\text{Supp}\,\psi\,,\ \eta\cdot a(v)=0\right\} = 0 \qquad \textit{for all}\quad \eta\in S^{N-1}$$

(resp.

(57) $$\text{meas}\left\{v\in\text{Supp}\,\psi\,,\ \tau+\eta\cdot a(v)=0\right\} = 0 \qquad \textit{for all}\quad (\eta,\tau)\in S^N \quad).$$

Then, $\int_{\mathbb{R}^k}\psi\,g_n\,dv$ is relatively compact in L^q_{loc} for all $q\in[1,p)$.

The preceding result allows in fact to recover all compactness results (for velocity averages) in the stability results given in the preceding section. Let us explain an additionnal technical idea : assume that f_n solves

(58) $$\frac{\partial f_n}{\partial t} + \xi\cdot\nabla_x f_n + \text{div}_\xi(F_n f_n) = C_n \qquad \text{in}\quad \mathbb{R}^{2N}_{x,\xi}\times(0,\infty)$$

where F_n is bounded in $L^1_{\rm loc}$, $C_n 1_{(|f_n|\le R)}$ is bounded in $L^1_{\rm loc}$ for all $R \in (0,\infty)$ and $\mathrm{div}_\xi\, F_n = 0$ on $\mathcal{D}'$. In fact, (58) holds in renormalized sense i.e. we have

$$\frac{\partial \beta(f_n)}{\partial t} + \mathrm{div}_x \{\beta(f_n)\} + \mathrm{div}_\xi \{F_n \beta(f_n)\} \;=\; \beta'(f_n) C_n \qquad \text{in} \quad \mathcal{D}' \tag{59}$$

for all $\beta \in C^1(\mathbf{R})$ such that β' is compactly supported on $\mathbf{R}$. We can also assume that (58) holds in the sense of distributions : in that case, we assume that C_n is bounded in $L^1_{\rm loc}$, f_n is bounded in $L^p_{\rm loc}$ for some $1 \le p \le \infty$, and F_n is bounded in $L^{p'}_{\rm loc}$. In the first case, we assume that f_n is relatively weakly compact in $L^1_{\rm loc}$.

We leave to the reader the verification of the fact that, in all the stability results - but for the (L) model - stated in the preceding sections, the above assumptions hold. And Theorem IV.2 immediately yields that, $\int_{\mathbf{R}^N} f_n \psi \, d\xi$ (in the second case), $\int_{\mathbf{R}^N} \beta_R(f_n)\psi \, d\xi$ (in the first case) are compact in $L^1_{\rm loc}(\mathbf{R}^{2N}_{x,\xi} \times (0,\infty))$ where $\beta_R = R\beta(\frac{t}{R})$, $R \in (0,\infty)$, $\beta \in C^1(\mathbf{R})$, $\beta' \equiv 1$ on $[-1,+1]$, $\beta(0) = 0$, $\beta'(t) \equiv 0$ if $|t| \ge 2$. The compactness of velocity averages stated in the previous section then follow easily from the bounds satisfied by f_n. In particular, the $f_n \log^+ f_n$ (for example) bound implies

$$|\beta_R(f_n) - f_n)| \;\le\; C\, f_n 1_{f_n \ge R} \;\le\; \frac{C}{\log R} f_n \log^+ f_n \qquad \text{for all} \quad R > 1\,.$$

And thus, $\beta_R(f_n)$ converges in L^1 to f_n as R goes to $+\infty$, uniformly in n.

In conclusion, we showed in this section how most of the compactness results stated in the preceding section can be deduced from velocity averages compactness results such as the ones stated above. In fact, only the "strong" L^1 or L^p convergences for (V) models or (L) models do not follow from the above results (at least immediately). The strong convergence results for (V) models follow in fact from results and methods explained in the next section. However, the strong convergence result for the (L) model uses velocity averaging together with new (and delicate) arguments.

Let us conclude this section with a few indications on the key ideas in the proof of Theorem III.7. Using "renormalized solutions techniques", one first obtains bounds of the following form : for all $R, T \in (0,\infty)$, there exists $C = C(R,T) \ge 0$ (independent of n) such that

$$\int_0^T dt \int_{\mathbf{R}^3} dx \int_{|\xi| \le R} \overline{a}^n_{ij} \, \frac{\partial \gamma(f_n)}{\partial \xi_i} \, \frac{\partial \gamma(f_n)}{\partial \xi_j} \, d\xi \;\le\; C \tag{60}$$

where $\overline{a}^n_{ij} = a_{ij} \underset{\xi}{*} f_n$ and $\gamma(t) = \frac{t}{1+t}$ on $[0,\infty)$. Then, using velocity averaging, one can show that

$$\overline{a}^n_{ij} \underset{n}{\to} \overline{a}_{ij} = a_{ij} \underset{\xi}{*} f \qquad \text{in} \quad L^1(\mathbf{R}^3_x \times B_R \times (0,T)) \quad \text{for all} \quad R, T \in (0,\infty) \tag{61}$$

$$(62) \qquad \begin{cases} \rho_n = \int_{\mathbf{R}^3} f_n \, d\xi \underset{n}{\rightarrow} \rho\left(= \int_{\mathbf{R}^3} f \, d\xi \right) \\ \text{in} \quad L^p(0,T; L^1(\mathbf{R}^3_x)) \quad \text{for all} \quad T \in (0,\infty) \,,\; 1 \le p < \infty \end{cases}$$

and

$$(63) \quad \gamma(f_n) \underset{\xi}{*} \kappa_\varepsilon \underset{n}{\rightarrow} \overline{\gamma} \underset{\xi}{*} \kappa_\varepsilon \qquad \text{in} \quad L^p(0,T; L^1(\mathbf{R}^6_{x,\xi})) \quad \text{for all } T \in (0,\infty) \,,\; 1 \le p < \infty$$

where $\overline{\gamma}$ is the weak limit in $L^1(0,T; L^1(\mathbf{R}^6_{x,\xi}))$ ($\forall\, T \in (0,\infty)$) of $\gamma(f_n)$ - extracting subsequences if necessary - and where $\kappa_\varepsilon = \frac{1}{\varepsilon^3}\kappa(\frac{\cdot}{\varepsilon})$, $\kappa \in C_0^\infty(\mathbf{R}^3)$, $\kappa \ge 0$, $\int_{\mathbf{R}^3} \kappa \, d\eta = 1$.

Next, we observe that we have on the set $\{(x,t) \in \mathbf{R}^3 \times (0,\infty) \,/\, \rho(x,t) > 0\}$

$$\sum_{i,j=1}^{3} \overline{a}_{ij}(x,\xi,t)\eta_i\eta_j > 0 \qquad \text{for all} \quad \xi \in \mathbf{R}^3 \,,\; \eta \in \mathbf{R}^3 \,,\; \eta \neq 0 \,.$$

Using (60)-(63) and this strict ellipticity on the set $\{\rho > 0\}$, one is able to show (see [28] for more details)

$$(64) \qquad \gamma(f_n) \underset{n}{\rightarrow} \overline{\gamma} \qquad \text{in} \quad L^p(0,T; L^1(\mathbf{R}^6_{x,\xi})) \quad \text{for all} \quad T \in (0,\infty) \,,\; 1 \le p < \infty \,.$$

Therefore, $\gamma(f_n)$ converges in measure on $\mathbf{R}^6_{x,\xi} \times (0,T)$ ($\forall\, T \in (0,\infty)$, and since γ is $1-1$ so does f_n. This convergence combined with the bounds available for f_n yields Theorem III.7.

In fact, it is worth mentioning that this method of proof can be used to prove compactness results for linear equations presenting hypoelliptic features but that are not hypoelliptic in a strict sense and we refer to [28] for some example in that direction.

V. Generalized flows.

An important tool in the analysis of (V) or (VB) models and in particular of the proofs of Theorems III.1 and III.5 is the possibility of being able to integrate the equation along the particle paths and thus to define in a meaningful way the particle paths and (or) the associated linear (first-order) transport equations. Notice that these particle paths are essentially defined by ordinary differential equations of the following form

$$(65) \qquad \dot{x} = \xi \quad , \quad \dot{\xi} = F(x,t)$$

and $F = -\Delta(V_0 \underset{x}{*} f)$. The system (65) is clearly a (non-autonomous) Hamiltonian system and classical ODE results à la Cauchy-Lipschitz would require F to be locally Lipschitz in x in order to solve (65) (and obtain a flow). However, it is quite clear from the bounds which are available that such a regularity is not known in the (VB) (or even (V)) models. Indeed, in the important example of a Coulomb interaction potential $V_0 = \frac{1}{|x|}$, such a regularity would imply (in particular) that ρ is locally bounded!

It turns out that less regularity is required if we are willing to accept to solve (65) for most initial conditions (a.e.) and to discard irrelevant initial conditions on a set of Lebesgue measure 0. This fact was discovered in R.J. DiPerna and P.L. Lions [**17**] and we want to review such results here. Let us immediately emphasize the fact that the Liouville invariance of the Lebesgue measure by flows associated to equations like (65) plays a crucial role in that analysis. Let us also mention that (65) may be studied from a Lagrange viewpoint where we insist on solving ordinary differential equations, or from an Euler viewpoint where we look at the following associated transport equation

$$\text{(66)} \qquad \frac{\partial f}{\partial t} + \operatorname{div}(uf) = 0 \quad \text{in } \mathbf{R}^N_y \times (0,\infty)$$

where $y = (x,\xi) \in \mathbf{R}^N$, $u = (\xi, F)$ satisfies

$$\text{(67)} \qquad \operatorname{div}_y(u) = 0 \quad \text{in } \mathcal{D}' .$$

In fact, the analysis introduced in [**17**] is valid for both viewpoints and the methods of proofs use in fact both sides of this coin. Il also applies to arbitrary nonconservative systems (66) i.e. arbitrary vectorfields u satisfying (67).

This is why, in order to give a flavour of these results, we shall first consider an autonomous situation

$$\text{(68)} \qquad \dot{Y} = u(Y) \quad , \quad t \in \mathbf{R}\,;\ Y(0,y) = y$$

where u satisfies (67). Then, we shall consider equations of the form (66) where $N = 6$, $y = (x,\xi) \in \mathbf{R}^3 \times \mathbf{R}^3$, $u = (\xi, F(x,t))$ that are thus directly related to (V) or (VB) models.

We begin with (68) and, in order to simplify the presentation, we assume that u admits a bounded invariant region say a smooth, bounded, connected open set Ω and

$$\text{(69)} \qquad u \cdot n = 0 \quad \text{on } \partial\Omega ,$$

where n is the unit outward normal to $\partial\Omega$.

Of course, (69) requires some regularity on u but since we are assuming that (67) holds in Ω, standard trace theorems give a perfect meaning to (69) when $u \in L^1(\Omega)$ for instance. Then, we would like to solve (68) for all $y \in \overline{\Omega}$ and obtain a flow $(t,y) \mapsto Y(t,y)$ on $\mathbf{R} \times \overline{\Omega}$ into $\overline{\Omega}$ that leaves the (restriction of the) Lebesgue measure invariant. But, since we want to cover situations where u has a limited regularity, we allow to consider generalized flows that we only define for almost all initial conditions $y \in \overline{\Omega}$. In order to do so, we define a Lebesgue-invariant a.e. flow Y solving (68) as a family of mappings

$(Y(t))_{t\in\mathbb{R}}$ from $\overline{\Omega}$ into $\overline{\Omega}$ such that : $Y \in C(\mathbb{R}_t; L^1(\Omega))$; $\lambda \circ Y(t) = Y(t)$ where λ is the restriction of Lebesgue measure to $\overline{\Omega}$;

$$Y(t+s) = Y(t) \circ Y(s) \qquad \text{a.e., for all} \quad t, s \in \mathbb{R} \tag{70}$$

and (68) holds for almost $(g,t) \in \Omega \times \mathbb{R}$. Let us recall that $\lambda \circ Y(t)$ is the measure (on $\overline{\Omega}$) defined by :

$$\int \varphi \, d(\lambda \circ Y(t)) = \int \varphi(Y(t,y))\, dy \quad , \quad \text{for all} \quad \varphi \in C(\overline{\Omega}) .$$

The following result can be deduced from [**17**]

Theorem V.1.

Let $u \in W^{1,1}(\Omega)$ satisfy (67) and (69).

(i) Then, there exists a unique Lebesgue-invariant a.e. flow Y solving (68). Furthermore, Y satisfies : $\dot{Y} \in C(\mathbb{R}_t; L^q(\Omega))$ and $Y \in L^q(\Omega; W^{1,q}(-T,T))$ ($\forall\, T \in (0,\infty)$) if $u \in L^q(\Omega)$ for $1 \le q < \infty$ (in particular we have always $\dot{Y} \in C(\mathbb{R}_t; L^{\frac{N}{N-1}}(\Omega))$, $Y \in L^{\frac{N}{N-1}}(\Omega; W^{1,\frac{N}{N-1}}(-T,T))$ ($\forall\, T \in (0,\infty)$) ; for almost all $y \in \Omega$, $Y \in C^1(\mathbb{R}_t)$, $u(Y) \in C(\mathbb{R}_t)$ and $\dot{Y} = u(Y)$.

(ii) Let $u_n \in W^{1,1}(\Omega)$ satisfy (67) and (69). Assume that u_n converges to u in $L^p(\Omega)$ for $1 \le p < \infty$. Let Y_n be the corresponding Lebesgue-invariant a.e. flow, then Y_n converges to Y in $C([-T,+T]; L^p(\Omega))$ ($\forall\, T \in (0,\infty)$).

(iii) Let $f_0 \in L^1(\Omega)$, then $f(y,t) = f_0(Y(t,y))$ is the unique renormalized solution of

$$\frac{\partial f}{\partial t} = u \cdot \nabla f \quad \text{in } \Omega \times \mathbb{R} \quad , \quad f \in C(\mathbb{R}_t; L^1(\Omega)) \quad , \quad f|_{t=0} = f_0 \, . \tag{71}$$

(iv) Under the conditions of (ii), let $f_0^n \in L^p(\Omega)$ for $1 \le p < \infty$ and let f^n be the renormalized solution of (71) with u replaced by u^n (so that by (iii) $f^n = f_0^n(Y_n(t,y))$). Then, if $f_0^n \underset{n}{\to} f_0$ in $L^p(\Omega)$, $f^n \underset{n}{\to} f = f_0(Y(t,y))$ in $C([-T,+T]; L^p(\Omega))$ ($\forall\, T \in (0,\infty)$).

Remark V.1.

Analogous results hold for non autonomous vectorfields replacing $u \in W^{1,1}(\Omega)$ by $u \in L^1(-T,T; W^{1,1}(\Omega))$ ($\forall\, T \in (0,\infty)$).

Remark V.2.

In [**17**], examples are given showing that it does not seem possible to relax in general the conditions upon u. □

We now give an application (and an adaptation of this result) to equations of the following form (for instance)

$$\frac{\partial f}{\partial t} + \xi \cdot \nabla_x f + F \cdot \nabla_\xi f = g \qquad \text{in} \quad \mathbb{R}^6_{x,\xi} \times \mathbb{R} \tag{72}$$

where we assume that $F = F(x,t)$ satisfies

(73) $$F \in L^1(-T,+T;L^{5/3}(\mathbf{R}^3)) + L^1(-T,+T;L^\infty(\mathbf{R}^3)) \quad (\forall\, T \in (0,\infty))\,,$$

(74) $$F \in L^1(\mathbf{R};\, W^{1,1}_{\text{loc}}(\Omega))\,,$$

and that g satisfies

(75) $$g \in L^1(-T,+T;L^1(\mathbf{R}^3)) + L^1(-T,+T;L^\infty(\mathbf{R}^3)) \quad (\forall\, T \in (0,\infty))\,.$$

Theorem V.2.

Let f_i be renormalized solutions of (72) such that $f_i|\xi|^2 \in L^\infty(\mathbf{R}_t; L^1(\mathbf{R}^6_{x,\xi}))$, $f_i \in C(\mathbf{R}_t; L^1 + L^q(\mathbf{R}^6_{x,\xi}))$ (for some $q \in [1,\infty)$) for $i = 1,2$. Then, if $f_1|_{t=0} \equiv f_2|_{t=0}$ a.e. on $\mathbf{R}^6_{x,\xi}$, $f_1 \equiv f_2$ a.e. on $\mathbf{R}^6_{x,\xi} \times \mathbf{R}$.

Remark V.3.

One can also obtain convergence results as in Theorem IV.1 above...

Remark V.4.

One can show extensions of this Theorem without requiring the $L^1(|\xi|^2)$ bound but where one assumes the existence of another solution f_0 of equation (72) such that $f_0 > 0$ a.e. and $f_0|\xi|^2 \in L^\infty(\mathbf{R}_t; L^1(\mathbf{R}^6_{x,\xi}))$. If we do not assume that $f_0 > 0$ a.e., we can simply conclude that $f_1 \equiv f_2$ on $\{f_0 > 0\}$... □

In order to prove Theorem V.2, one first observes that, by the arguments of **[17]**, one can prove that $f = |f_1 - f_2|$ is a renormalized solution of (72) with $g = 0$ and which vanishes at $t = 0$. It is thus enough to show that if $f \in C(\mathbf{R}_t; L^1_{x,\xi}) \cap L^\infty_{x,\xi,t}$ solves (72) (in renormalized sense or in distributions sense) with $g = 0$ and if $f \geq 0$ on $\mathbf{R}^6_{x,\xi} \times \mathbf{R}_t$, $f|_{t=0} \equiv 0$ then $f \equiv 0$. This fact is established if we show that $\iint_{\mathbf{R}^6} f\,dx\,d\xi$ is independent of t. To this end, we consider $\varphi \in C_0^\infty(\mathbf{R}^3)$, $\varphi \equiv 1$ on B_1, Supp $\varphi \subset B_2$ and we multiply the equation by $\varphi(\frac{x}{n})\,\varphi(\frac{\xi}{n})$ and integrate by parts over $\mathbf{R}^6_{x,\xi}$. We then find for all $t \geq 0$ (say)

$$\begin{aligned}
&\iint_{|x|,|\xi|\leq n} f(t)\,dx\,d\xi \\
&\leq \int_0^t \iint_{\mathbf{R}^6} \Big|\varphi\big(\frac{\xi}{n}\big)\frac{\xi}{n}\cdot\nabla\varphi\big(\frac{x}{n}\big) + F(x,t)\cdot\frac{1}{n}\nabla\varphi\big(\frac{\xi}{n}\big)\,\varphi\big(\frac{x}{n}\big)\Big| f(s)\,ds\,dx\,d\xi \\
&\leq C\int_0^t ds \int_{n\leq|x|\leq 2n}\int_{\mathbf{R}^3} |f(s)|\,dx\,d\xi \\
&\qquad + C\int_0^t ds \int_{\mathbf{R}^3} |F(x,s)|\,dx\,\frac{1}{n}\int_{n\leq|\xi|\leq 2n} f(x,\xi,s)\,d\xi\,.
\end{aligned}$$

The first integral clearly goes to 0 as n goes to $+\infty$. The reason why the second one also goes to 0, allowing us to conclude, is that we have

$$\begin{aligned}\frac{1}{n}\int_{n\leq|\xi|\leq 2n} f(x,\xi,s)\,d\xi &\leq \frac{C}{n}\left[\int_{n\leq|\xi|\leq 2n} |f|^{5/2}(x,\xi,s)\,d\xi\right]^{2/5} (n^3)^{3/5}\\ &\leq \frac{C}{n} n^{9/5}\,\|f\|^{3/5}_{L^\infty_{x,\xi}}\, n^{-4/5}\left(\int_{n\leq|\xi|\leq 2n} f|\xi|^2\,d\xi\right)^{2/5}\\ &\leq C\left(\int_{n\leq|\xi|\leq 2n} f|\xi|^2\,d\xi\right) \quad (\in L^\infty(\mathbf{R}_t;L^{5/2}(\mathbf{R}^3_x)))\,.\end{aligned}$$

In order to apply these results to (V) or (VB) models, one thus has to check that $V \in L^1_{\mathrm{loc}}(\mathbf{R}_t; W^{1,1}_{\mathrm{loc}}(\mathbf{R}^3_x))$. In fact, one can check using (37) that $V \in L^\infty(\mathbf{R}_t; W^{1,1}_{\mathrm{loc}}(\mathbf{R}^3_x))$. Indeed, in view of (37), it is enough to check that

$$\sup_{t\geq 0}\int_{\mathbf{R}^3} \rho(t)\log\rho(t)\,dx < \infty$$

or

$$\sup_{t\geq 0}\int_{\mathbf{R}^3} \rho(t)\log^+\rho(t)\,dx < \infty\,.$$

This fact is a consequence of the following refinement of Lemma II.1.

Lemma V.1.

Let $f \in L^1(\mathbf{R}^{2N}_{x,\xi})$ satisfy : $f \geq 0$, $f|\xi|^2 \in L^1$, $f\log^+ f \in L^1$ then $\rho(x) = \int_{\mathbf{R}^N} f\,d\xi$ satisfies : $\rho\log^+\rho \in L^1$. In addition if $f|x|^2 \in L^1$, $f\,|\log f| \in L^1$ then $\rho\log\rho \in L^1$.

Remark V.5.

In fact if $f(1+|\xi|^2)$, $f\log^+ f$ are bounded in L^1, the proof below shows that $\rho\log^+\rho$ is bounded in L^1.

Proof.

The second part of Lemma V.1 follows from the first part using the argument shown in section II. The first part requires several steps.

First of all, we write $f = f\,1_{(f<1)} + f\,1_{(f\geq 1)}$ and we observe it is enough to show that both $\rho' = \int f\,1_{(f<1)})\,d\xi$ and $\overline{\rho} = \int f\,1_{(f\geq 1)}\,d\xi$ satisfy the "$L^1\log^+ L^1$" integrability. Indeed, Lemma IV.1 then follows by observing that we have for all $a, b \geq 0$

$$(a+b)\log^+(a+b) \leq (a+b)\log^+[2\max(a,b)] \leq (\log 2)(a+b) + 2a\log^+ a + 2b\log^+ b\,.$$

Then, $\rho'\log^+\rho' \in L^1$ since $f\,1_{(f<1)}$ and $|\xi|^2 f\,1_{(f<1)} \in L^1_{x,\xi}$ so that by Lemma II.1, $\rho' \in L^{\frac{N+2}{N}}(\mathbf{R}^N_x)$ and thus $\rho'\log^+\rho' \in L^1(\mathbf{R}^N_x)$ observe that $t\log^+ t \leq Ct^{\frac{N+2}{N}}$ for all $t \geq 0$, for some $C > 0$).

Next, we study $\overline{\rho}$. To this end, we set $A = \{(x,\xi)\ /\ f \geq 1\}$ and $A_x = \{\xi\ /\ f(x,\xi) \geq 1\}$ for $x \in \mathbf{R}^N$. Clearly, $1_A \in L^\infty \cap L^1(\mathbf{R}^{2N}_{x,\xi})$ and $|\xi|^2\,1_A \in L^1(\mathbf{R}^{2N}_{x,\xi})$ since $\iint_A |\xi|^2\,dx\,d\xi \leq \iint f|\xi|^2\,dx\,d\xi < \infty$. Therefore, applying again Lemma II.1, we deduce that $|A_x|$

$(= \int 1_A \, d\xi) \in L^{\frac{N+2}{N}}(\mathbb{R}_x^N)\,(\cap L^1(\mathbb{R}_x^N))$.

Then, we set $\Phi(t) = t \log^+ t$: a convex function on $\mathbb{R}$. Therefore, by Jensen's inequality, we have on $\mathbb{R}_x^N$

$$\begin{aligned}\Phi(\overline{\rho}) &= \Phi\left(|A_x|\left[\int_{A_x} f \, d\xi |A_x|^{-1}\right]\right) \le |A_x|^{-1} \int_{A_x} \Phi(|A_x| f) \, d\xi \\ &\le \int_{A_x} f \log^+(|A_x| f) \, d\xi \le \int_{A_x} f \left[\log^+ f + \log^+ |A_x|\right] d\xi \\ &\le \int f \log^+ f \, d\xi + \overline{\rho} \log^+ |A_x| \,.\end{aligned}$$

By assumption, the first term on the right-hand side belongs to $L^1(\mathbb{R}_x^N)$. In order to estimate the second term, we first remark that we have for all $a, b \ge 0$

$$ab \le e^{\frac{N+2}{N}} 1_{a>0} + \frac{N}{N+2} b \log^+ b \,.$$

And we finally deduce

$$\Phi(\overline{\rho}) \le \int f \log^+ f \, d\xi + |A_x|^{\frac{N+2}{N}} 1_{|A_x|>1} + \frac{N}{N+2} \Phi(\overline{\rho})$$

and we can conclude. □

References.

[1] V.I. Agoshkov, *Spaces of functions with differential-difference characteristics and smoothness of solutions of the transport equation.* Soviet Math. Dokl, 29 (1984), p. 662-666.

[2] L. Arkeryd, *On the Boltzmann equation.* Arch. Rat. Mech. Anal., 45 (1972), p. 1-34.

[3] C. Bardos, F. Golse and D. Levermore, *Fluid dynamic limits of kinetic equations. Convergence proofs for the Boltzmann equation.* Preprint, announced in C.R. Acad. Sci. Paris, 309 (1989), p. 727-732.

[4] J. Bergh and J. Löfstrom, Interpolation spaces. An introduction. Springer, Berlin, 1976.

[5] L. Boltzmann, *Weitere studien über das wärme gleichgenicht unfer gasmoleküler.* Sitzungsberichte der Akademie der Wissenschaften, Wien, 66 (1872), p. 275-370. Translation : *Further studies on the thermal equilibrium of gaz molecules*, 88-174; In Kinetic Theory 2, ed. S.G. Brush, Pergamon, Oxford, 1966.

[6] T. Carleman, Problèmes mathématiques dans la théorie cinétique des gaz. Notes rédigées par L. Carleson et O. Frostman, Publications mathématiques de l'Institut Mittag-Leffler, Almquist and Wikselles, Uppsala, 1957.

[7] C. Cercignani, The Boltzman equation and its applications. Springer, Berlin, 1988.

[8] S. Chapman and T.G. Cowling, The mathematical theory of non-uniform gases. Cambridge Univ. Press, 1939 ; 2nd edit., 1952.

[9] P. Degond and B. Lucquin-Desreux, *The Fokker-Planck asymptotics of the Boltzmann collision operator in the Coulomb case.* Preprint.

[10] L. Desvillettes, *On an asymptotics of the Boltzman equation when the collisions become grazing.* Preprint.

[11] R.J. DiPerna and P.L. Lions, *On the Cauchy problem for Boltzmann equations : Global existence and weak stability.* Ann. Math., 130 (1989), p.321-366.

[12] R.J. DiPerna and P.L. Lions, *Global solutions of Boltzmann's equation and the Entropy inequality.* Arch. Rat. Mech. Anal., 114 (1991), p. 47-55.

[13] R.J. DiPerna and P.L. Lions, *On the Fokker-Planck-Boltzmann equation.* Comm. Math. Phys., 120 (1988), p. 1-23.

[14] R.J. DiPerna and P.L. Lions, *Solutions globales d'quations du type Vlasov-Poisson.* C.R. Acad. Sci. Paris, 307 (1988), p. 655-658.

[15] R.J. DiPerna and P.L. Lions, *Global weak solutions of Vlasov-Maxwell equations.* Comm. Pure Appl. Math., XLII (1989), p. 729-757.

[16] R.J. DiPerna and P.L. Lions, *Global weak solutions of kinetic equations.* Sem. Matematico Torino, 46 (1988), p. 259-288.

[17] R.J. DiPerna and P.L. Lions, *Ordinary differential equations, Sobolev spaces and transport theory.* Invent. Math., 98 (1989), p. 511-547.

[18] R.J. DiPerna, P.L. Lions and Y. Meyer, *L^p regularity of velocity averages.* Ann. I.H.P. Anal. Non Lin., 8 (1991), p. 271-287.

[19] P. Gérard, P.L. Lions and T. Paul, work in preparation.

[20] R. Glassey, personnal communication.

[21] F. Golse, P.L. Lions, B. Perthame and R. Sentis, *Regularity of the moments of the solution of a transport equation.* J. Funct. Anal., 76 (1988), p. 110-125.

[22] F. Golse, B. Perthame and R. Sentis, *Un résultat de compacité pour les équations de transport et applications au calcul de la limite de la valeur propre principale d'un opérateur de transport.* C.R. Acad. Sci. Paris, 301 (1985), p. 341-344.

[23] H. Grad, *Principles of the kinetic theory of gases,* in Flügge's Handbuch der Physik XII, Springer, Berlin, (1958), p. 205-294.

[24] R. Hamdache, *Existence in the large and asymptotic behaviour for the Boltzmann equation.* Japan J. Appl. Math., 2 (1985), p. 1-15.

[25] R. Illner and M. Shinbrot, *The Boltzmann equation. Global existence for a rare gas in an infinite vacuum.* Comm. Math. Phys., 95 (1984), p. 217-226.

[26] E.M. Lifschitz and L.P. Pitaevski, Physical kinetics. Pergamon, Oxford, 1951.

[27] P.L. Lions, *On kinetic equations.* In Proceedings of the International Congress of Mathematicians, Kyoto, 1990. Springer, Tokyo, 1991.

[28] P.L. Lions, *On Boltzman and Landau equations.* Preprint.

[29] P.L. Lions and T. Paul, *Sur les mesures de Wigner.* Preprint.

[30] P.L. Lions and B. Perthame, *Propagation of moments and regularity for the 3-dimensional Vlasov-Poisson system.* Invent. Math., 105 (1991), p. 415-430.

[31] P.L. Lions, B. Perthame and E. Tadmor, *A kinetic formulation of multidimensional scalar conservation laws and related equations.* Preprint.

[32] J.C. Maxwell, *On the dynamical theory of gases.* Phil. Trans. Roy. Soc. London, 157 (1866), p. 49-88.

[33] T. Nishida and K. Imaï, *Global solutions to the initial value problem for the non-linear Boltzmann equation.* Publ. R.I.M.S. Kyoto Univ., 12 (1976), p. 229-239.

[34] K. Pfaffelmoser, *Global classical solutions of the Vlasov-Poisson system in three dimensions for general initial data.* J. Diff. Eq., 95 (1992), p. 281-303.

[35] J. Schaeffer, *Global existence of smooth solutions to the Vlasov-Poisson system in three dimensions.* Preprint.

[36] C. Truesdell and R.G. Muncaster, Fundamentals of Maxwell's kinetic theory of a simple monoatomic gas. Academic Press, New-York, 1963.

[37] S. Ukai, *Solutions of the Boltzmann equation.* In: Patterns and Waves. Qualitative analysis of differential equations. North-Holland, Amsterdam, (1986), p. 37-96.

Kinetic Models for Semiconductors

Peter A. Markowich[1]

The goal of these lecture notes, which originate from the seven lectures given by the author at the short course, is to discuss various topics of present interest in the area of kinetic models for semiconductors. Obviously this is a very large and expanding area (particularly due to the ongoing miniaturisation of semiconductor devices) and, thus, a complete presentation (of sufficent mathematical rigor) would by large exceed the scope of these notes. Thus a choice of the topics had to be made. Sections 1 and 2 are concerned with quantum mathematical models (used for ultra-integrated high frequncy devices) and Section 3 with the hydrodynamic (Euler-Poisson) model, which in the near future may very well replace the by now standardly used drift-diffusion equations in real life device simulations.

The choice of these topics is based on the fact that these models entered only recently in applications and that their analysis is rather novel and mathematically challenging.

Certainly, there are other topics in the research area, which also satisfy the above mentioned criteria, e.g. the semi-classical Boltzmann equation of solid state physics and its mean free path limits. Although these issues were discussed in the lectures they were not included in these notes since a rather recent overview [5] can be found in the literature and since by now, its analysis is rather well understood.

At this point I want to mention that the research for Section 1 (time dependent quantum mechanical model) was done in collaboration with N. Mauser and the research for Section 3 (hydrodynamic semiconductor models) in collaboration with U. Ascher, P. Pietra and C. Schmeiser.

[1]Fachbereich Mathematik, TU Berlin, Straße des 17 Juni 136, W-1000 Berlin 12, Germany

1. The Classical Limit of a Time-Dependent Quantum-Vlasov Equation

We consider the quantum mechanical motion of a particle ensemble in $I\!R^3$. The pure state wave functions of a one-particle approximation satisfy the Schrödinger equations

$$i\varepsilon\frac{\partial}{\partial t}\psi_j^\varepsilon = -\frac{\varepsilon^2}{2}\Delta\psi_j^\varepsilon + V^\varepsilon\psi_j^\varepsilon, x \in I\!R^3, t > 0, j \in N \tag{1.1}$$

$$\psi_j^\varepsilon(t=0) = \varphi_j^\varepsilon, x \in I\!R^3, j \in N \tag{1.2}$$

and the (mixed) position density n^ε and current density J^ε are given by:

$$n^\varepsilon(x,t) := \sum_{j=1}^{\infty} \lambda_j^\varepsilon |\psi_j^\varepsilon(x,t)|^2, x \in I\!R^3, t > 0 \tag{1.3}$$

$$J^\varepsilon(x,t) := \varepsilon \sum_{j=1}^{\infty} \lambda_j^\varepsilon Im(\overline{\psi_j^\varepsilon}\nabla_x\psi_j^\varepsilon(x,t)), x \in I\!R^3, t > 0.$$

The coefficients $\lambda_j^\varepsilon \geq 0$ are the occupation probabilities of the $L^2(I\!R^3)$-orthonormed initial states φ_j^ε and $\varepsilon > 0$ is the (normalized) Planck constant. The effective mass was set to 1.

We model the potential self-consistently:

$$(a) V^\varepsilon = H^\varepsilon *_x n^\varepsilon or (b) V^\varepsilon = -H^\varepsilon *_x n^\varepsilon \tag{1.4}$$

where the two-body interaction potential is given by

$$(a) H^\varepsilon(x) = \frac{1}{4\pi|x|} *_x G^{2\varepsilon} \tag{1.5}$$

with

$$G^\delta = \frac{1}{(\pi\delta)^{\frac{3}{2}}} e^{(-\frac{|x|^2}{\delta})} \quad \text{for} \quad \delta > 0. \tag{1.6}$$

(1.4)(a) corresponds to the repulsive case (electrons) and (1.4)(b) to the attractive case (see [1] for a physical example).

Obviously, H^ε is a smoothed version of the Green's function $\frac{1}{4\pi|x|}$ of the operator $-\Delta$ on $I\!R^3$ and, thus, V^ε is a smoothed Coulomb potential associated with the density n^ε. We shall see later on that the choice (1.6) of the smoother G^δ allows to introduce a main feature into the analysis of the self-consistent (Schrödinger) problem, which somewhat simplifies the passage to the limit $\varepsilon \to 0$.

We define the mixed state density matrix

$$z^\varepsilon(r,s,t) := \sum_{j=1}^{\infty} \lambda_j^\varepsilon \overline{\psi_j^\varepsilon}(r,t)\psi_j^\varepsilon(s,t), \quad r,s \in I\!R^3, t > 0 \tag{1.7}$$

and its Wigner transformation

$$w^\varepsilon(x,v,t) := \frac{1}{2\pi^3}\int_{\mathbb{R}^3} z^\varepsilon(x+\frac{\varepsilon}{2}\eta, x-\frac{\varepsilon}{2}\eta, t)e^{iv\eta}d\eta. \tag{1.8}$$

Obviously, w^ε is the η-Fouriertransform of the density matrix z^ε after the change of the coordinates

$$(r,s) \leftrightarrow (x,\eta), r = x+\frac{\varepsilon}{2}\eta, s = x-\frac{\varepsilon}{2}\eta \tag{1.9}$$

(see [1]-[5]). Then w^ε satisfies the so-called Wigner equation

$$w_t^\varepsilon + v\cdot\nabla_x w^\varepsilon + \theta^\varepsilon[V^\varepsilon]w^\varepsilon = 0; x,v\in \mathbb{R}^3, t>0 \tag{1.10}$$

$$w^\varepsilon(t=0) = w_I^\varepsilon = 0; x,v\in\mathbb{R}^3, \tag{1.11}$$

where $\theta^\varepsilon[V^\varepsilon]$ is the pseudo-differential operator

$$\theta^\varepsilon[V^\varepsilon]w^\varepsilon = \frac{1}{(2\pi)^3}\int_{\mathbb{R}^3_{v'}}\int_{\mathbb{R}^3_\eta} \delta^\varepsilon[V^\varepsilon](x,\eta,t)w^\varepsilon(x,v',t)e^{i(v-v')\cdot\eta}d\eta dv'. \tag{1.12}$$

The symbol $\delta^\varepsilon[V^\varepsilon]$ is given by

$$\delta^\varepsilon[V^\varepsilon](x,\eta,t) = i\frac{V^\varepsilon(x+\frac{\varepsilon}{2}\eta,t) - V^\varepsilon(x-\frac{\varepsilon}{2}\eta,t)}{\varepsilon}. \tag{1.13}$$

The initial datum w_I^ε is the Wigner transform of the initial density matrix

$$z_I^\varepsilon(r,s) = \sum_{j=1}^\infty \lambda_j^\varepsilon \overline{\psi_j^\varepsilon}(r)\psi_j^\varepsilon(s), \quad r,s\in\mathbb{R}^3. \tag{1.14}$$

A simple (formal) calculation shows that the macroscopic densities n^ε and J^ε, defined by (1.3), can (formally) be written as

$$n^\varepsilon = \int_{\mathbb{R}^3_v} w^\varepsilon dv, J^\varepsilon = \int_{\mathbb{R}^3_v} vw^\varepsilon dv. \tag{1.15}$$

The main advantage of the Wigner transform lies in the structural similarity of the quantum-mechanical Wigner equation (1.9) to the classical Vlasov equation which is obtained formally in the limit $\varepsilon\to 0$. We shall now present prerequisites for the analysis and then carry out the limit $\varepsilon\to 0$ rigorously (for details see [6]).

We shall use the following assumptions on the initial density matrix z_I^ε:
(A1) for $\varepsilon\in(0,\varepsilon_0], \varepsilon>0$ fixed, we have

(1) $\lambda_j^\varepsilon \geq 0 \quad \forall_j\in N, \{\psi_j^\varepsilon\}_{j\in N}$ is orthonormal in $L^2(\mathbb{R}^3)$
and
(2) there are $\gamma=\gamma(\varepsilon)>0, A=A(\varepsilon)>0$ such that

$$\sum_{j=1}^\infty \lambda_j^\varepsilon \|\psi_j^\varepsilon\|^2_{H^{\frac{5+1}{2}}(\mathbb{R}^3)} \leq A$$

and
(A2) there is $C>0$ independent of $\varepsilon\in(0,\varepsilon_0]$ such that

$$\sum_{j=1}^\infty \lambda_j^\varepsilon + \varepsilon^2\sum_{j=1}^\infty \lambda_j^\varepsilon\|\nabla\psi_j^\varepsilon\|^2_{L^2(\mathbb{R}^3)} + \frac{1}{\varepsilon^3}\sum_{j=1}^\infty(\lambda_j^\varepsilon)^2 \leq C$$

for $\varepsilon \in (0, \varepsilon_0]$.

The assumptions (A1), (A2) can easily be restated in terms of the (initial) density operator Z_I^ε (with integral kernel z_I^ε

$$Z_I^\varepsilon : L^2(\mathbb{R}^3) \leftarrow L^2(\mathbb{R}^3), (Z_I^\varepsilon f)(s) = \int_{\mathbb{R}^3_r} z_I^\varepsilon(r,s) f(r) dr$$

(see [6] for details).

In [2] it was shown that the Schrödinger–Poisson problem has a global regular solution in the repulsive case if (A1), (A2), hold. Exactly the same techniques give the existence of a unique global regular solution of the regularized Schrödinger–Poisson problem (1.1)–(1.6) in the repulsive case. The attractive case can be dealt with using the methods of [1].Using standard regularity arguments the formal calculations which lead to (1.15) can be justified and it can be shown that the local conservation law

$$n_t^\varepsilon + div_x J^\varepsilon = 0 \quad in D'(\mathbb{R}^3 \times (0,\infty)) \tag{1.16}$$

is valid. Multiplication of (1.10) by $|v|^2$ and integration give the conservation of energy (cf. [1][2])

$$\int_{\mathbb{R}^3_x}\int_{\mathbb{R}^3_v} |v|^2 w^\varepsilon(x,v,t) dv dx \pm \int_{\mathbb{R}^3_v} |\nabla W^\varepsilon(x,t)|^2 dx \tag{1.17}$$
$$= \int_{\mathbb{R}^3_x}\int_{\mathbb{R}^3_v} |v|^2 w_I^\varepsilon(x,v) dv dx \pm \int_{\mathbb{R}^3_x} |\nabla W_I^\varepsilon(x)|^2 dx, \quad \forall t > 0,$$

with the auxilary potential

$$W^\varepsilon = \pm H^{\frac{\varepsilon}{2}} *_x n^\varepsilon \tag{1.18}$$

which we introduce because of the smoothing in V^ε. The "+" signs in (1.17), (1.18) hold for the repulsive case and the "–" signs for the attractive case. Another main ingredient is furnished by the introduction of a smoothed Wigner function, the so-called Husimi-function $w^{\varepsilon,\delta}$ defined by

$$w^{\varepsilon,\delta} := w^\varepsilon *_x G^\delta(x) *_v G^\delta(v) \tag{1.19}$$

The Husimi-function of a pure state is known to be nonnegative if the condition $\delta \geq \varepsilon$ (reflecting the uncertainity principle) holds (see [3],[4]). We can show (for mixed states)

LEMMA 1.1. Let (A1),(A2) hold. Then
(1) $w^{\varepsilon,\delta} \geq 0$ on $\mathbb{R}^3_x \times \mathbb{R}^3_v \times (0,\infty)$ for $\delta \geq \varepsilon$
(2) Let w^I, w be accumulation points of w_I^ε and w^ε, respectively, in the $L^2(\mathbb{R}^3_x \times \mathbb{R}^3_v)$ -weak and, resp., $L^\infty((0,\infty), L^2(\mathbb{R}^3_x \times \mathbb{R}^3_v))$- weak* topologies. Then $w_I \geq 0$ a.e. on $\mathbb{R}^3_x \times \mathbb{R}^3_v$ and $w \geq 0$ a.e. on $\mathbb{R}^3_x \times \mathbb{R}^3_v \times (0,\infty)$.

Note that

$$n^{\varepsilon,\delta} := \int_{\mathbb{R}^3_v} w^{\varepsilon,\delta} dv = n^\varepsilon *_x G^\delta \tag{1.20}$$

holds. The regularized Coulomb potential V^ε is the Coulomb potential of the Husimi-density $n^{\varepsilon,2\varepsilon}$. Analogously, the auxilary potential W^ε defined by (1.18) is the Coulomb potential of the Husimi density $n^{\varepsilon,\varepsilon}$:

$$W^\varepsilon = \pm \frac{1}{4\pi|x|} *_x n^{\varepsilon,\varepsilon} \tag{1.21}$$

Also, we shall use the classical interpolation estimates [7]:

LEMMA 1.2
Let

$$u \in L^2(\mathbb{R}^3_x \times \mathbb{R}^3_v), |v|^2 u \in L^1(\mathbb{R}^3_x \times \mathbb{R}^3_x)$$

Define

$$\rho := \int_{\mathbb{R}^3_v} u dv, J := \int_{\mathbb{R}^3_v} v u dv$$

Then

$$\rho \in L^{\frac{7}{5}}(\mathbb{R}^3), \quad J \in L^{\frac{7}{6}}(\mathbb{R}^3_x)$$

and there exists a constant $c > 0$ independent of u such that

$$||\rho||_{L^{\frac{7}{5}}(\mathbb{R}^3_x)} \leq c||u||^{\frac{4}{7}}_{L^2(\mathbb{R}^3_x \times \mathbb{R}^3_v)} |||v|^2 u||^{\frac{3}{7}}_{L^1(\mathbb{R}^3_x \times \mathbb{R}^3_v)} \tag{1.22}$$

$$||J||_{L^{\frac{7}{6}}(\mathbb{R}^3_x)} \leq c||u||^{\frac{2}{7}}_{L^2(\mathbb{R}^3_x \times \mathbb{R}^3_v)} |||v|^2 u||^{\frac{5}{7}}_{L^1(\mathbb{R}^3_x \times \mathbb{R}^3_v)}. \tag{1.23}$$

Using (A1),(A2) we conclude by integrating (1.10) over $\mathbb{R}^3_x \times \mathbb{R}^3_v$

$$0 \leq \int_{\mathbb{R}^3_x} \int_{\mathbb{R}^3_v} w^\epsilon_I(x,v) dv dx = \int_{\mathbb{R}^3_x} \int_{\mathbb{R}^3_v} w^\epsilon_I(x,v,t) dv dx \leq C, t \geq 0, \tag{1.24}$$

$$||n^\epsilon||_{L^\infty((0,\infty));L^1(\mathbb{R}^3_x)} \leq C \tag{1.25}$$

$$||w^\epsilon_I||_{L^2(\mathbb{R}^3_x \times \mathbb{R}^3_v)} \leq C; ||w^\epsilon_I||_{L^\infty((0,\infty));L^2(\mathbb{R}^3_x \times \mathbb{R}^3_v)} \leq C \tag{1.26}$$

Here and in the sequel we denote by C not necessarily equal constants independent of ε and t. By means of (1.24)-(1.26), Lemmata 1.1 and 1.2 and the conservation laws (1.16), we can derive an uniform estimate for the kinetic energy which readily yields for the Husimi density:

$$||n^{\epsilon,2\epsilon}||_{L^\infty((0,\infty);L^{\frac{7}{5}})} \leq C, \tag{1.27}$$

$$||J^{\epsilon,2\epsilon}||_{L^\infty((0,\infty);L^{\frac{7}{6}})} \leq C, \tag{1.28}$$

and

$$||V^\epsilon||_{L^\infty((0,\infty);L^6(\mathbb{R}^3_x))} + ||\nabla V^\epsilon||_{L^\infty((0,\infty);L^2(\mathbb{R}^3_x))} \tag{1.29}$$

follows by estimating (1.4).

Using compactness we have, after restriction to a subsequence if necessary,

$$w^\epsilon_I \xrightarrow{\varepsilon \to 0} w_I \quad \text{in} L^2(\mathbb{R}^3_x \times \mathbb{R}^3_v) \quad \text{weakly} \tag{1.30}$$

$$w \xrightarrow{\varepsilon \to 0} w \quad \text{in} L^\infty((0,\infty); L^2(\mathbb{R}^3_x \times \mathbb{R}^3_v)) \quad \text{weak*} \tag{1.31}$$

$$V \xrightarrow{\varepsilon \to 0} V^\epsilon \quad \text{in} L^\infty((0,\infty); L^2(\mathbb{R}^3_x)) \quad \text{weak*} \tag{1.32}$$

$$\nabla V^\epsilon \xrightarrow{\varepsilon \to 0} \nabla V^\epsilon \quad \text{in} L^\infty((0,\infty); L^6(\mathbb{R}^3_x)) \quad \text{weak*} \tag{1.33}$$

$$n^{\epsilon,2\epsilon} \xrightarrow{\varepsilon \to 0} n \quad \text{in} L^\infty((0,\infty); L^{\frac{7}{5}}(\mathbb{R}^3_x)) \quad \text{weak*} \tag{1.34}$$

$$J^{\epsilon,2\epsilon} \xrightarrow{\varepsilon \to 0} J \quad \text{in} L^\infty((0,\infty); L^{\frac{7}{6}}(\mathbb{R}^3_x)) \quad \text{weak*} \tag{1.35}$$

Note that the crucial role of the Husimi density $n^{\epsilon,2\epsilon}$ lies in bypassing n for which we only have an L^1 estimate. Identifying the limit we easily see that V is the Newtononian potential of n. The uniform L^1 - estimates on n^ε as well as on the kinetic energy estimate yield:

LEMMA 1.3
(1)

$$n^\varepsilon \xrightarrow{\varepsilon\to 0} n,\ J^\varepsilon \xrightarrow{\varepsilon\to 0} J \quad \text{in} D'(\mathbb{R}^3_x \times (0,\infty))$$

(2)

$$n = \int_{\mathbb{R}^3_v} w dv,\ J = \int_{\mathbb{R}^3_v} v w dv$$

In preparation to carry out the limit $\varepsilon \to 0$ in the Wigner equation (1.10) we prove

LEMMA 1.4
The symbol $\delta^\varepsilon[V^\varepsilon]$ of $\theta^\varepsilon[V^\varepsilon]$ can be written as

$$\delta^\varepsilon[V^\varepsilon](x,\eta,t) = F^\varepsilon(x,\eta,t) + i\nabla_x V^\varepsilon(x,t)\cdot\eta \tag{1.36}$$

where F^ε satisfies the estimate

$$||F^\varepsilon(.,.,t)||_{L^2(B_R\times B_R)} \le C_{\sigma,q}(R)\varepsilon^\sigma|\nabla V^\varepsilon(.,t)|_{W^{\sigma,q}(B_{2R})} \tag{1.37}$$

for every $R > 0, 0 < \sigma < 1$, and $2 \le q < \infty$. Here B_R denotes the ball in $\mathbb{R}^3$ with radius R and center in the origin.

The imbedding $W^{2,7/5}(B_{2R}) \hookrightarrow W^{1+\sigma,q}(B_{2R})$ with $\frac{3}{q} - \sigma = \frac{8}{7}$ and the standard localization argument for the Poisson equation on $\mathbb{R}^3_x$ (using $||\nabla V^\varepsilon(\cdot,t)||_{L^2(\mathbb{R}^3_x)} \le C$) together with Lemma 1.4 and (1.27) yield

$$||F^\varepsilon||_{L^\infty((0,\infty);L^2(B_R\times B_R))} \le C(R)\varepsilon^{\frac{5}{14}} \tag{1.38}$$

(with $q = 2$, $\sigma = 5/14$).

Now let $\varphi = \varphi(x,v,t)$ be a C^∞–test function such that the support of $F_v\varphi$ (Fouriertransformation with respect to v) is compact in $\mathbb{R}^3_x \times \mathbb{R}^3_\eta \times [0,\infty)$. We multiply the Wigner equation (1.10) by φ, integrate by parts and use Plancherel's theorem. Then

$$\int w^\varepsilon(\varphi_t + v\cdot\nabla_x\varphi)dxdvdt - \int F^\varepsilon(\overline{F_v\varphi})dxd\eta dt \tag{1.39}$$
$$+ \int w^\varepsilon\nabla_x V^\varepsilon\cdot\nabla_v\varphi dxdvdt = -\int w_I^\varepsilon\varphi dxdv$$

follows. Obviously, the passage to the limit can be carried out in the first two integrals on the left hand side of (1.39) and in the integral on the right hand side. In order to be able to pass to the limit in the term involving (the electric field) ∇V^ε, we reiterate that

$$||V^\varepsilon||_{L^\infty((0,\infty);W^{2,7/5}(B_R))} \le C(R) \tag{1.40}$$

for every $R > 0$. Also, from the local conservation law (1.16) rewritten for the Husimi quantities we obtain via integration by parts and the Sobolev inequality

$$||\nabla V^\varepsilon_t||_{L^\infty((0,\infty);W^{-1,21/11}(B_R))} \le C(R) \tag{1.41}$$

With (1.40) and (1.41) and a standard compactness result (Cor.3.4 in [7]) then imply that for every $R > 0$ and for every $T > 0$ there is a subsequence such that

$$V^\varepsilon \xrightarrow{\varepsilon\to 0} V \quad \text{in} \quad C([0,T];H^1(B_R)) \quad \text{strongly} \tag{1.42}$$

Thus, we can pass to the limit in the third integral on the left hand side of (1.39). We collect the results in

THEOREM 1.1
Let the assumptions (A1),(A2) hold. Let w_I be an accumulation point of w_I^ε in the $L^2(\mathbb{R}_x^3\times\mathbb{R}_v^3)$ weak topology as $\varepsilon\to 0$. Then there exists a sequence $(w^\varepsilon, V^\varepsilon, n^\varepsilon, J^\varepsilon)$ of solutions of the regularized Wigner-Poisson problem (1.10)-(1.15),(1.4)-(1.6) such that(1.30)-(1.35) hold and further

$$n^\varepsilon \overset{\varepsilon\to 0}{\longrightarrow} n \quad \text{in} D'(\mathbb{R}_x^3\times(0,\infty))$$

$$J^\varepsilon \overset{\varepsilon\to 0}{\longrightarrow} J \quad \text{in} D'(\mathbb{R}_x^3\times(0,\infty))$$

where (w, V, n, J) is a distributional solution of the Vlasov–Poisson equation

$$w_t + v\cdot\nabla_x w - \nabla_x V\cdot\nabla_v w = 0, x, v\in\mathbb{R}^3, t>0$$

$$w(t=0) = w_I, x, v\in\mathbb{R}^3$$

$$n = \int_{\mathbb{R}_v^3} w dv, J = \int_{\mathbb{R}_v^3} vw dv, x\in\mathbb{R}^3, t>0$$

$$(a) V = \frac{1}{4\pi}\int_{\mathbb{R}_y^3}\frac{n(y,t)}{|x-y|}dy or (b) V = -\frac{1}{4\pi}\int_{\mathbb{R}_y^3}\frac{n(y,t)}{|x-y|}dy$$

The solution satisfies the macroscopic conservation law

$$n_t + div_x J = 0 \quad \text{in} \quad D'(\mathbb{R}^3\times(0,\infty)),$$

has globally bounded energy

$$\int_{\mathbb{R}_x^3}\int_{\mathbb{R}_v^3}|v|^2 w(x,vt)dvdx + \int_{\mathbb{R}_x^3}|\nabla_x V(x,t)|^2 dx \le C, t\ge 0$$

and is globally bounded in $L^1(\mathbb{R}_x^3\times\mathbb{R}_v^3)\cap L^2(\mathbb{R}_x^3\times\mathbb{R}_v^3)$.

We shall now construct an admissible (in particular nonnegative definite) sequence of density matrices for every appropriate phase-space density such that a corresponding sequence of Wigner transforms converges to w_I as $\varepsilon\to 0$.

LEMMA 1.5
Let $F, G\in C^\infty(\mathbb{R}^3)$, F complex valued, G real valued and nonnegative on $\mathbb{R}^3$. Define

$$z^\varepsilon(r,s) = \overline{F}(r)F(s)\tilde{G}\left(\frac{r-s}{\varepsilon}\right), r, s\in\mathbb{R}^3, \tag{1.43}$$

where $\tilde{G}$ denotes the Fourier transform of G. Let w^ε be the Wigner transform of z^ε and set

$$w(x,v) = |F(x)|^2)G(v). \tag{1.44}$$

Then the corresponding density operator Z^ε (with kernel z^ε) is hermitian, nonnegative definite and the following relations hold:

$$tr(Z^\varepsilon) = \frac{1}{(2\pi)^{\frac{3}{2}}}\int_{\mathbb{R}_x^3}\int_{\mathbb{R}_v^3} w(x,v)dvdx \tag{1.45}$$

$$\varepsilon^2 tr(-\Delta Z^\varepsilon) = \frac{1}{(2\pi)^{\frac{3}{2}}}\int_{\mathbb{R}_x^3}\int_{\mathbb{R}_v^3}|v|^2 w(x,v)dvdx \tag{1.46}$$

$$+\varepsilon^2 \int_{\mathbb{R}^3} |\nabla F(x)|^2 dx \int_{\mathbb{R}^3_v} G(v)dv$$

$$\frac{1}{\varepsilon^3} tr((Z^\varepsilon)^2) \xrightarrow{\varepsilon \to 0} \frac{1}{4} \int_{\mathbb{R}^3_x} \int_{\mathbb{R}^3_v} w(x,v)^2 dv dx \tag{1.47}$$

$$w^\varepsilon \xrightarrow{\varepsilon \to 0} w \quad \text{in} L^2(\mathbb{R}^3_x \times \mathbb{R}^3_v) \quad \text{strongly} \tag{1.48}$$

Using this Lemma we can prove the following theorem (see [6] for details).

THEOREM 1.2
Let $w_I \in L^1(\mathbb{R}^3_x \times \mathbb{R}^3_v) \cap L^2(\mathbb{R}^3_x \times \mathbb{R}^3_v)$, $|v|^2 w_I \in L^1(\mathbb{R}^3_x \times \mathbb{R}^3_v)$, $w_I \geq 0$ a.e. on $\mathbb{R}^3_x \times \mathbb{R}^3_v$. Then there exists a sequence $\varepsilon_m \downarrow 0$ as $m \to \infty$ and a sequence $\{z^{\varepsilon_m}_{I,m}\}_{m \in N}$ of density matrices, whose eigenfunction expansion satisfies (A1 and (A2) (with C independent of ε_m and m) such that the Wigner transforms $w^{\varepsilon_m}_{I,m}$ of the kernels $z^{\varepsilon_m}_{I,m}(r,s)$ converge to w_I in $L^2(\mathbb{R}^3_x \times \mathbb{R}^3_v)$ strongly as $m \to \infty$.

We remark that an analysis of the classical limit of the Wigner-Poisson problem (without the smoothing (1.5)) of the self-consistent potential V^ε was recently performed by P.L. Lions and T. Paul [9].

Quantum Steady States

We consider an ensemble of electrons moving in $\mathbb{R}^3$ under the influence of their self-generated Coulomb forces and on external real-valued potential V_e. In the quantum mechanical steady state case the orthonormal pure state wave functions u_j satisfy the eigenvalue problem for the Schrödenger equation:

$$-\frac{\varepsilon^2}{2}\Delta u_j + (V + V_e)u_j = \lambda_j u_j, x \in \mathbb{R}^3, j \in N \tag{2.1}$$

where V denotes the (self-consistent) Coulomb potential, λ_j the eigenvalue (energy-value) of the j-th pure state and ε the scaled Planck constant.

Assume now that a positive decreasing distribution function ϕ of the eigenvalues is given. Then the (mixed state) electron position density can be written as

$$n(x) = \sum_{j=1}^{\infty} \phi(\lambda_j - \lambda_F) u_j(x)^2 \tag{2.2}$$

(note that the eigenvalues λ_j are real and the eigenfuction can be chosen to be real-valued). Here λ_F denotes the Fermi-energy, which is usually determinded by a normalisation condition on the total charge density.

$$\int_{\mathbb{R}^3} n(x)dx = 1 \Rightarrow \sum_{j=1}^{\infty} \phi(\lambda_j - \lambda_F) = 1 \tag{2.3}$$

The Coulomb potential V then reads

$$V(x) = \frac{1}{4\pi|x|} * n. \tag{2.4}$$

Before going into the analysis of the nonlinear eigenvalue problem (2.1)-(2.4) we shall take a look at its classical mechanics equivalent. Then the phase space distribution function $F = F(x, v)$ solves the steady state Vlasov- Poisson equation:

$$v\nabla_x F - \nabla_x(V + V_e) \cdot \nabla_v F = 0, \quad x, v \in \mathbb{R}^3 \tag{2.5}$$

$$n(x) = \int_{\mathbb{R}^3} F(x, v)dv \tag{2.6}$$

with the Newtonian potential V given by (2.4). Since any function of the total energy $\frac{1}{2}|v|^2 + V + V_e$ solves (2.5), we set

$$F(x, v) = \phi(\frac{1}{2}|v|^2 + V + V_e - \lambda_F) \tag{2.7}$$

where ϕ is the given energy distribution and again λ_F the Fermi energy determined by the normalisation

$$\int_{\mathbb{R}^3} n(x)dx = 1 \tag{2.8}$$

or equivalently

$$\int_{\mathbb{R}^3_x}\int_{\mathbb{R}^3_v}\phi(\frac{1}{2}|v|^2+V+V_e-\lambda_F)dvdx=1$$

Thus, in the clasiscal mechanics case the Newtonian solution of the Poisson-equation

$$-\Delta V=\int_{\mathbb{R}^3}\phi(\frac{1}{2}|v|^2+V+V_e-\lambda_F)dv\quad(=n) \tag{2.9}$$

with λ_F given by (2.8), has to be sought.

Of particular physical importance is the case of the Boltzmann distribution

$$\phi(t)=e^{\beta^* t} \tag{2.10}$$

where (up to a scaling factor) β^* is the reciprocal of the given (fixed) ambient temperature $T>0$. Then the Poisson equation(2.9) reads

$$-\Delta V=\frac{n_e(x)e^{-\beta^* V}}{\int_{\mathbb{R}^3}n_e(x)e^{-\beta^* V(x)}dx} \tag{2.11}$$

with the "external density".

$$n_e=e^{-\beta^* V_e} \tag{2.12}$$

In the sequel we shall point out main analytical features of the quantum problem (2.1)-(2.4) in the Boltzmann distributed case. In particular we shall see that the mean field equation (2.11) turns out to be the classical limit of (2.1)-(2.4).
The analysis of the steady state Schrödinger-Poisson problem (2.1)-(2.4) can be done by three different (though intrinsically related) techniques.
a) a fixed point argument on the potential V applied directly to (2.1)-(2.4),
b) a minimization method (see[10],[11]) similar to the one used in [12],[13] for the analysis of the classical mean-field equation (2.11),
c) transformation to the so called Bloch-equation [14],[15].

Here we shall pursue the approach c),since in particular it allows a convenient analysis of the classical limit $\varepsilon\to 0$ by employing the Wigner transformation. We shall only present a sketch of the analysis, details can be found in [14],[15].

For the Boltzmann distributed case (2.10) we define the unnnormalized density matrix

$$z(r,s,\beta)=\sum_{j=1}^{\infty}e^{-\lambda_j\beta}u_j(r)u_j(s) \tag{2.13}$$

for $0<\beta\le\beta^*$. Formally we have

$$z=\exp(-\frac{1}{2}(H_r[V]+H_s[V])\beta)\delta(r-s) \tag{2.14}$$

where $H[V]$ denotes the Hamiltonian (depending on the Coulomb potential V):

$$H[V]=-\frac{\varepsilon^2}{2}\Delta+V_e+V. \tag{2.15}$$

$H_r[V]$ and $H_s[V]$ denote copies of $H[V]$ acting on the r and s variables respectively.

Equivalently, z is the solution of the parabolic IVP (also called Bloch-equation)

$$z_\beta = \frac{\varepsilon^2}{4}\Delta_{(r,s)}z + \frac{1}{2}(V_e(r) + V_e(s) + V(r) + V(s)) \quad z, r, s \in I\!R^3, 0 < \beta \leq \beta^* \tag{2.16}$$

$$z(r, s, \beta = 0) = \delta(r - s). \tag{2.17}$$

The Coulomb potential has to be determined by the nonlinear coupling

$$V = \frac{1}{4\pi|x|} * n \tag{2.18}$$

with the density

$$n(x) = z(x, x, \beta^*) / \int_{I\!R^3} z(x, x, \beta^*)dx \tag{2.19}$$

(taking into account the normalization (2.3)). Note that the coupling of (2.16) with (2.18),(2.19) occurs only at $\beta = \beta^*$ ie: the problem (2.16)- (2.19) is not a standard evolution problem .

The basic assumption for the analysis is

(A1) $V_e \in L^\infty_{loc}(I\!R^3), \exists \aleph \in I\!R : V_e \geq \aleph$ in $I\!R^3$,
$V_e \to \infty$ as $|x| \to \infty, \exists \alpha \in (0,1) : e^{-\alpha\beta^* V_e} \in L^1(I\!R^3)$

It guarantees that the spectrum of $H_e = -\frac{\varepsilon^2}{2}\Delta + V_e$ is discrete and that $e^{(-\beta H_e)}$is of trace class for $\frac{\beta}{\beta^*} \geq \alpha$.

A somewhat natural approach to the existence proof of (2.16)- (2.19)is a fixed point argument on the Coulomb potential V. The main ingredient is to establish an $L^\infty(I\!R^3)$ bound of V, which is a rather technical issue (see [15]) . Then the spectral properties of H_e carry over to $H[V]$ and the equivalence of the Schrödinger-Poisson problem (2.1)-(2.4) and the Bloch-Poisson problem (2.16)-(2.19) is easily established.
Uniqueness can be proved by a monotonicity method, based on the inequality [10],[11],[14]:

$$\int_{I\!R^3} (n[V_1] - n[V_2])(V_1 - V_2)dx \geq 0 \tag{2.20}$$

where $n[W]$ denotes the electron density obtained from the potential W (in place of the Coulomb potential V). (2.20) reflects the tendency of electrons to occupy states of low potential energy. Together with the Poisson equation $-\Delta V = n$ it implies uniqueness immediately. We state the result in

THEOREM 2.1 Let (A1) hold. Then there exists a unique triple $(V, n, \{\lambda_j,\}_{j\in N})$ and a complete orthonormal system $\{u_j\}_{j\in N}$ of $L^2(I\!R^3)$ functions such that $(V, n, \{\lambda_j\}_{j\in N}, \{u_j\}_{j\in N})$ is a solution of the Schrödinger Poisson problem (2.1)-(2.4) in the case of the Boltzmann distribution (2.10).

Results for general distributions (including the physically significant Fermi-Dirac case) in a slightly different setting can be found in [11].

To understand the classical limit $\varepsilon \to 0$ we introduce the coordinate transformation (first step of the Wigner transform,see Section 1)

$$(r, s) \leftrightarrow (x, \eta), r = x + \frac{\varepsilon}{2}\eta, s = x - \frac{\varepsilon}{2}\eta, \tag{2.21}$$

use superscripts ε for the dependent variables and set

$$y^\varepsilon(x,\eta,\beta,\beta^*) = \frac{z^\varepsilon(x+\frac{\varepsilon}{2}\eta, x-\frac{\varepsilon}{2},\beta)}{\int_{\mathbb{R}^3} z^\varepsilon(x,x,\beta^*)dx} \tag{2.22}$$

for $(x,\eta) \in \mathbb{R}^6, 0 \le \beta \le \beta^*$. Then y^ε solves

$$y^\varepsilon{}_\beta = \frac{\varepsilon^2}{8}\Delta_x y^\varepsilon + \frac{1}{2}\Delta_\eta y^\varepsilon - \frac{1}{2}(V_e(x+\frac{\varepsilon}{2}\eta) \tag{2.23}$$

$$+V_e(x-\frac{\varepsilon}{2}\eta) + V^\varepsilon(x+\frac{\varepsilon}{2}\eta) + V^\varepsilon(x-\frac{\varepsilon}{2}\eta)y^\varepsilon,$$

$$y^\varepsilon(x,\eta,\beta=0,\beta^*) = \frac{\delta(\eta)}{\varepsilon^3\int_{\mathbb{R}^3} z^\varepsilon(x,x,\beta^*)dx} \tag{2.24}$$

with

$$n^\varepsilon = y^\varepsilon(x,\eta=0,\beta=\beta^*,\beta^*), -\Delta V^\varepsilon = n^\varepsilon \tag{2.25}$$

It was shown in [15] that

$$\varepsilon^3\int_{\mathbb{R}^3} z^\varepsilon(x,x,\beta^*)dx \to A(\beta^*) > 0 \tag{2.26}$$

Then the formal limit of (2.23),(2.24),(2.25) reads (using the superscripts 0 for the limit quantities)

$$y_\beta{}^0 = \frac{1}{2}\Delta_\eta y + (V_e(x) + V^0(x))y^0, (x,\eta) \in \mathbb{R}^6, 0 < \beta \le \beta^* \tag{2.27}$$

$$y^0(x,\eta,\beta=0,\beta^*) = \frac{\delta(\eta)}{A(B^*)} \tag{2.28}$$

$$n^0 = y^0(x,\eta,\beta=\beta^*,\beta^*), -\Delta V^0 = n^0 \tag{2.29}$$

We denote the Fourier transformations of y^0 with respect to η by $Y^0(x,v,\beta=\beta^*,\beta^*)$ and obtain from (2.27),(2.28)

$$Y_\beta{}^0 = \frac{1}{2}|v|^2Y^0 + (V_e(x) + V^0(x))y^0, (x,v) \in \mathbb{R}^6, 0 < \beta \le \beta^* \tag{2.30}$$

$$Y^0(x,v,\beta=0,\beta^*) = \frac{1}{(2\pi)^{\frac{3}{2}}A(\beta^*)} \tag{2.31}$$

(second step of the Wigner transform, see Section 1). Solving the O.D.E. gives

$$Y^0(x,v,\beta=0,\beta^*) = \frac{1}{(2\pi)^{\frac{3}{2}}A(\beta^*)}e^{\beta(-\frac{1}{2}|v|^2+V_e(x)+V^0(x)} \tag{2.32}$$

Evaluating at $\beta = \beta^*$, taking into account (2.29) and the normalisation condition (2.3) gives the limit density

$$n^0(x) = \frac{n_e(x)e^{-\beta^*V^0}}{\int_{\mathbb{R}^3} n_e(x)e^{-\beta^*V^0(x)}dx} \tag{2.33}$$

and the Coulomb potential

$$V^0(x) = \frac{1}{4\pi|x|} * \frac{n_e e^{-\beta^*V^0}}{\int_{\mathbb{R}^3} n_e e^{-\beta^*V^0}dx} \tag{2.34}$$

given by the mean field equation. The uniform estimates necessary to make the formal limit process presented above rigorous, can be found in [15].

To conclude this section we mention an interesting, so far open, related problem. We assumed that the thermodynamic parameters $\int_{R^3} n dx$(total charge) and ambient temperature $= \frac{1}{\beta^*}$ are given. In the time evolution (quantum Boltzmann) case it is reasonable to assume that the total charge and energy are conserved (determined by the initial state of the system). Therefore, it is reasonable to prescribe the total charge and the total energy (instead of the temperature). As already pointed out in Section 1, the total energy is given by

$$\mathcal{E} = \frac{\varepsilon^2}{2} \frac{\sum_{j=1}^{\infty} e^{-\beta^* \lambda_j} \|\nabla u_j\|^2_{L^2(R^3)}}{\sum_{j=1}^{\infty} e^{-\beta^* \lambda_j}} + \int_{R^3} (V_e + V) n dx \tag{2.35}$$

A simple calulation using the Schrödinger equation (2.1) gives

$$\mathcal{E} = \frac{\sum_{j=1}^{\infty} e^{-\beta^* \lambda_j} \lambda_j}{\sum_{j=1}^{\infty} e^{-\beta^* \lambda_j}} \tag{2.36}$$

Thus instead of a-priorily fixing β^*,we can prescribe the value of $\mathcal{E}(= \infty$ say) and use (2.36) to determine β^*. The Schrödinger Poisson problem then reads

$$-\frac{\varepsilon^2}{2} \Delta u_j + (V_e + V) u_j = \lambda_j u_j, j \in N$$

$$n = \frac{\sum_{j=1}^{\infty} e^{-\beta^* \lambda_j} u_j^2}{\sum_{j=1}^{\infty} e^{-\beta^* \lambda_j}}$$

$$1 = \frac{\sum_{j=1}^{\infty} e^{-\beta^* \lambda_j} \lambda_j}{\sum_{j=1}^{\infty} e^{-\beta^* \lambda_j}}$$

$$V = \frac{1}{4\pi |x|} * n$$

We remark that the classical analogue of this problem has been investigated in [13].

3. Phase-Plane Analysis of Hydrodynamic Models

A scaled version of the one-dimensional Euler-Poisson (or hydrodynamic) model for a collisionless gas of negatively charged particles (electron plasma) in a vacuum reads:

$$n_t + (nu)_x = 0 \tag{3.1}$$

$$(nu)_t + (nu^2 + p) = nE \tag{3.2}$$

$$\left[n\left(\frac{u^2}{2} + \frac{3}{2}T\right)\right]_t + \left[nu\left(\frac{u^2}{2} + \frac{3}{2}T\right) + pu\right]_x = nuE \tag{3.3}$$

(see, e.g., [24]). Here $n > 0$ denotes the particle density, u the average particle velocity, p the pressure of the gas, $T > 0$, the particle temperature, and E the (negative) electric field, which is generated by the Coulomb force of the particles:

$$E_x = n - C(x). \tag{3.4}$$

The function $C = C(x) > 0$ stands for the density of fixed, positively charged background ions.

The equations (3.1)-(3.3) represent mass, momentum, energy balance, respectively (as derived from the semiconductor Boltzmann equation[9]). As equation of state we use the ideal gas law

$$p = nT\ . \tag{3.5}$$

and the (scaled) entropy of the gas is given by

$$S = ln\frac{p}{n^{\frac{5}{3}}} = ln\frac{T}{n^{\frac{2}{3}}}. \tag{3.6}$$

The particle current density is

$$J = nu. \tag{3.7}$$

Here we only consider the steady state case $n_t = (nu)_t = \left[n\left(\frac{u^2}{2} + \frac{3}{2}T\right)\right]_t = 0$. In particular, we want to model n^+nn^+ (three-layer) structures, i.e. the background ion-density C is supposed to be large in the two outer n^+ - layers and of moderate size, e.g. $C \equiv 1$, in the inner n-region (see [22]). Then with the inner n-region being located in the x interval $[0, \beta]$, the problem can be written as

$$\left(\frac{J^2}{n} + nT\right)_x = nE, \quad 0 < x < \beta, \tag{3.8}$$

$$\left(\frac{J^2}{2n^2} + \frac{5}{2}T\right)_x = E, \quad 0 < x < \beta, \tag{3.9}$$

$$E_x = n - 1, \quad 0 < x < \beta. \tag{3.10}$$

where J is a constant. The influence of the outer n^+-layers is modelled by the boundary conditions

$$n(0) = n(\beta) = \overline{n}, \tag{3.11}$$

where $\bar{n}(>> 1)$ represents the value of the background ion density C in the n^+ regions. In the following, we consider the current-controlled case, i.e. we assume that J is prescribed. Also we prescribe the temperature T at the upstream boundary. If, for a given J, the triple $(n(x), T(x), E(x))$ is a solution of (3.8)-(3.11), then the triple $(n(\beta - x), T(\beta - x), E(\beta - x))$ is a solution for $-J$. Thus, it suffices to consider the case $J \geq 0$ and the temperature boundary condition reads

$$T(0) = T_0 \tag{3.12}$$

The soundspeed of the flow given by (n, T) is defined by

$$c(n, T) := \sqrt{\frac{5}{3}T} \tag{3.13}$$

Then, the flow is subsonic, if

$$|u| < c(n, T) \Leftrightarrow J^2 < \frac{5}{3}n^2 T \tag{3.14}$$

and it is supersonic if

$$|u| > c(n, T) \Leftrightarrow J^2 > \frac{5}{3}n^2 T. \tag{3.15}$$

An interesting and practically relevant situation is to assume subsonic flow in the outer n^+ -regions. Then if the ion density $C(x) = 1$ (in the inner n-region) corresponds to a supersonic flow, transonic solutions (n, T) may be discontinuous. Therefore we have to supplement (3.8)-(3.12) by Rankine-Hugeniot (jump)-conditions for the one- sided limits of the electron density and of the temperature at the shock:

$$\frac{J^2}{2n_l^2} + \frac{5}{2}T_l = \frac{J^2}{2n_r^2} + \frac{5}{2}T_r \tag{3.16}$$

$$\frac{J^2}{n_l} + n_l T_l = \frac{J^2}{n_l} + n_r T_r. \tag{3.17}$$

Obviously, the field E is in $C[0, \beta]$, if, e.g., $n \in L^1(0, \beta)$.

Also, in order to select physically relevant weak solutions, we have to impose the so called entropy condition (see[25]). A way of stating it is to require that a shock occurs from a supersonic to a subsonic state (in the direction of the flow, which is the x direction since $J > 0$), i.e.

$$\frac{5}{3}n_l^2 T_l < J^2 < \frac{5}{3}n_r^2 T_r \tag{3.18}$$

In the following, two states (n_l, T_l, E) and (n_r, T_r, E) will be called *conjugate* when (3.16)-(3.18) holds.

If the energy balance equation is replaced by the assumption of isentropic flow, then from (3.6) the following constitutive equation is obtained for the pressure:

$$p(n) = kn^{\frac{5}{3}} \tag{3.19}$$

where k is a positive constant. Using (3.5) and (3.19), the system reads in the isentropic case:

$$(\frac{J^2}{n} + kn^{\frac{5}{3}})_x = nE, \quad 0 < x < \beta \tag{3.20}$$

$$E_x = n - 1, \quad 0 < x < \beta \tag{3.21}$$

$$n(0) = n(\beta) = \overline{n} \tag{3.22}$$

The temperature T is a given function of the density n

$$T = kn^{\frac{5}{3}} \tag{3.23}$$

and the soundspeed reduces to $c(n) = \sqrt{\frac{5}{3}kn^{\frac{2}{3}}}$.

In the sequel we shall refer to the non-isentropic problem (3.8)-(3.12),(3.16)-(3.18) as (NISP), while the isentropic problem (3.20)-(3.23) (with the jump condition $\frac{J^2}{n_l} + kn_l^{\frac{5}{3}} = \frac{J^2}{n_r} + kn_r^{\frac{5}{3}}$) will be referred to as (ISP).

The existence of subsonic solutions of the isentropic model is discussed in [3]. These results have been extended to multidimensional potential flows in [18]. The existence of one-dimensional transonic solutions was obtained in [20], [21] by means of a viscosity method and in [17] by means of a phase plane analysis. In all these papers collision effects can be included in the isentropic case. The non-isentropic model has been investigated in [23], where existence of transonic solutions was shown by extending the phase-plane analysis.

Here we present the main ideas of the phase-plane analysis for the problems under consideration. Details can be found in [17] and [23]. The main result is the existence of a continuum of solutions of the problems (ISP) and (NISP), containing at least one solution for every positive value of β. We shall see that the non-isentropic model (NISP) admits solutions, which are structurally different from those in the isentropic solution-continuum. In particular, the solution continuum of (NISP) contains solutions with two shocks (under certain conditions on the data), while the isentropic solutions of (ISP) have at the most one shock.

Numerical computations illustrating the possible solution structures are reported. For this purpose we use regularizations of the problems, adding a second order (viscous) term to the equation (3.8) in the non-isentropic case (to (3.20) in the isentropic case). The regularized problem was then solved by the general purpose boundary value ODE code COLSYS [16].

At first we construct a set of solutions (n, E, β) of (ISP), which contains at least one solution $(n, E,)$ for every $\beta > 0$ and which forms a continuum (i.e., a closed and connected subset) in an appropriately chosen function space.

We denote

$$\sigma = \frac{5}{3J^2}k \tag{3.24}$$

for $k > 0$ and define $n_c(\sigma) := \frac{1}{\sigma^{3/8}}$. Using (3.13) and (3.23), a simple computation shows that the line $n = n_c(\sigma)$ is sonic.

The problem (ISP) reads

$$J^2 F(n;\sigma)_x = nE, \quad 0 < x < \beta \tag{3.25}$$

$$E_x = n - 1, \quad 0 < x < \beta \tag{3.26}$$

$$n(0) = n(\beta) = \overline{n} \tag{3.27}$$

where F is given by

$$F(n;\sigma) = \frac{1}{n} + \frac{3}{5}\sigma n^{\frac{5}{3}}$$

The jump and entropy conditions at a point are

$$\frac{1}{n_l} + \frac{3}{5}\sigma n_l^{\frac{5}{3}} = \frac{1}{n_r} + \frac{3}{5}\sigma n_r^{\frac{5}{3}} \tag{3.28}$$

$$n_l < n_c(\sigma) < n_r \tag{3.29}$$

The assumption of subsonic boundary conditions implies $\overline{n} > n_c(\sigma)$ and the assumption of supersonic flow for the background ion density $C(x) = 1$ implies $n_c(\sigma) > 1$, i.e. $0 < \sigma < 1$.
The system (3.25),(3.26) is integrable and a first integral is given by:

$$G(n;\sigma) = -\frac{(n-1)^2}{2n^2} + \frac{3}{5}\sigma n_l^{\frac{2}{3}}(n - \frac{5}{2}) \tag{3.30}$$

The function G is constant along trajectories and vanishes at the stationary point $(n, E) = (1,0)$. Another point of interest is the so- called critical point $(n, E) = (n_c(\sigma), 0)$. The derivative $\frac{\partial F(n;\sigma)}{\partial n}$ vanishes at $n_c(\sigma)$. The sonic line $n = n_c(\sigma)$ splits the phase plane of into the subsonic domain $n > n_c(\sigma)$ and the supersonic domain $0 < n < n_c(\sigma)$.
The phase portrait of (3.25),(3.26) is shown in Fig. 1

The stationary point (1,0) is a center. The critical point $(n_c(\sigma), 0)$ is a point of non-uniqueness, through which the so-called critical trajectory, denoted by $T_c(\sigma)$, passes twice, namely once on its way from the subsonic into the supersonic domain and once on its way back from the supersonic into the subsonic domain. ($T_c(\sigma)$ is marked by the thickly drawn curve in Fig. 1). In the sequel we shall use the notation

$$T_c^{sub}(\sigma) = T_c(\sigma) \cap \{(n, E) | n \geq n_c(\sigma), E \in \mathbb{R}\}$$

$$T_c^{sup}(\sigma) = T_c(\sigma) \cap \{(n, E) | n < n \leq n_c(\sigma), E \in \mathbb{R}\}$$

Note that all points inside $T_c^{sup}(\sigma)$ lie on a periodic orbit about (1,0).

No trajectory except $T_c(\sigma)$ passes from the subsonic into the supersonic domain or vice versa. For every $E \neq 0$ there are exactly two trajectories, one defined in the subsonic domain and the other in the supersonic domain, which terminate ($E < 0$), or , respectively, initiate ($E > 0$) at the point $(n_c(\sigma), E)$ of the sonic line. The sonic line is always reached (and departed from) at finite values of x.
Now let (n_0, E_0) be a subsonic point, which lies above the trajectory segment $T_c^{sub}(\sigma) \cap \{(n, E) | E < 0\}$. Then the trajectory segment starting at (n_0, E_0) intersects every line $n = \tilde{n}$ with $\tilde{n} > 0$ exactly once, attaining a positive E-value at the point of intersection.

We denote by $T_c(\sigma)$ the image of the supersonic loop $T_c^{sup}(\sigma)$ under the conjugation map defined by the condition (3.28) ($T_c(\sigma)$ is the dashed curve in Fig. 1).

Now we consider the problem (3.25)-(3.29) with β as a parameter and construct a continuum of solutions. The construction starts with the point $(\overline{n}, 0)$ considered as a solution for $\beta = 0$. Obviously, this point can be considered as 'minimal' trajectory of a continuum of subsonic orbits starting and ending on the line $n = \overline{n}$. The 'maximal' subsonic trajectory consists of the two segments of $T_c^{sub}(\sigma)$ between $n = n_c(\sigma)$ and $n = \overline{n}$. The β-value corresponding to this solution is denoted by $2a_c$. Fig 2 shows a typical subsonic electron density n.

For $\beta > 2a_c$ transonic solutions must be constructed. Since jumps are only allowed from the supersonic to the subsonic region, the transition from a subsonic to supersonic state must be smooth. Therefore transonic solutions must follow initially the critical trajectory starting at a point $(\overline{n}, E_1) \in T^c(\sigma)$ with $E_1 < 0$. At some point a shock occurs and the right state (obeying the jump condition) lies on the conjugate curve $T^c(\sigma)$. Then a subsonic trajectory starting on $T^c(\sigma)$ is followed back to $n = \overline{n}$. In Fig.3 a typical transonic density n is presented. The value of E at the shock is positive, since the function n is increasing after the jump. Fig. 4 shows a transonic solution with the shock at a point in the lower half phase-plane $E < 0$.

Under the assumption $\overline{n} > n^*(\sigma)$, where $n^*(\sigma)$ denotes the maximal n-value on the conjugate curve $T^c(\sigma)$, a subsonic trajectory leading to $n = \overline{n}$ starts out at every point on $T^c(\sigma)$. Therefore, the construction of the solutions continues as follows. After all the points on $T_c^{sup}(\sigma)$ have been "used" as jump locations we arrive at a smooth solution by following $T_c^{sup}(\sigma)$ back into the subsonic region to $n = \overline{n}$. The β-value corresponding to this solution is denoted by $2a_c + 2b_c$.

For $\beta > 2a_c + 2b_c$ the construction can be repeated periodically. The solutions perform k loops along $T_c^{sup}(\sigma)$ and then jump back to the subsonic region, except for the values $\beta_n = 2a_c + 2kb_c, k = 1, 2...$, corresponding to continuous transonic solutions. Fig.5 shows the electron density for a typical solution with $\beta_1 < \beta < \beta_2$.

Now denote by (n_e, E_e) a solution (n, E) of (ISP) (for the given value of β)extended to the x-interval $(0, \infty)$ by setting

$$(n_e(x), E_e(x)) = \begin{cases} (n(x)E(x)), & 0 < x < \beta \\ (0, E(\beta)), & x > \beta \end{cases} \tag{3.31}$$

Simple continuity arguments show that β varies continuously along the above constructed solution set and that the set of extended solutions (n_e, E_e, β) forms a continuum in the $L^q(0, \infty) \times C[0, \infty] \times \mathbb{R}_0^+$-topology for every $1 \leq q \leq \infty$.
We collect the results in

THEOREM 3.1
Let the assumptions $0 < \sigma < 1$ and $\overline{n} \geq n^*(\sigma)$ hold. Then there exists a set S of solutions (n, E, β) of (ISP), containing a solution (n, E) for every $\beta \geq 0$, such that the set S_e of extended solutions (n_e, E_e, β) forms a continuous curve in the $L^q(0, \infty) \times C[0, \infty] \times \mathbb{R}_0^+$-topology for every $1 \leq q \leq \infty$. this curve of solutions consists of a bounded subset of subsonic solutions and of an unbounded subcontinuum of transonic solutions. Countably many transonic solutions are smooth for $x \in [0, \beta]$ and all others have exactly one shock in $[0, \beta]$.

The case $\overline{n} < n^*(\sigma)$ can be analyzed in a similar way, but it requires that the downstream boundary condition $n(\beta) = \overline{n}$ is 'relaxed' for certain values of β in order to allow for boundary jumps [21]. We refer to [17] for a complete discussion of this case.

It is clear that the value $\frac{5}{3}$ in the constitutive equation (3.19) for the pressure can be replaced by any $\gamma \geq 1$. The mathematical analysis presented above is not affected by such a change.

We now turn here to the analysis of solutions of (NISP) which will be presented piecewise as trajectories of isentropic systems of the form (3.25),(3.26). Indeed, assume that a solution (n, T, E) of (3.9) is smooth in the interval $(x_0, x_1) \subseteq [0, \beta]$. Then a simple calculation shows that

$$T = kn^{\frac{2}{3}} \quad \text{in}(x_0, x_1), \tag{3.32}$$

holds, i.e. the entropy S is constant on every interval of regularity. In particular, if the interval of regularity is $[0, x_0), x_0 > 0$, then

$$T = \frac{T_0}{\overline{n}^{\frac{2}{3}}} n^{\frac{2}{3}} \quad \text{in}[0, x_0), \tag{3.33}$$

We now use $k = \frac{T_0}{\overline{n}^{\frac{2}{3}}}$ in (3.24) and denote

$$\sigma_0 = \frac{5}{3J^2} \frac{T_0}{\overline{n}^{\frac{2}{3}}} \tag{3.34}$$

Then (3.8)-(3.10) reduces to (3.25)-(3.26) (with σ replaced by σ_0) in $[0, x_0)$, i.e., the solution (n, E) is (at least initially) a trajectory of (3.25),(3.26) starting on the line $n = \overline{n}$ and the temperature T is given by (3.33). Also here we assume that the line $n = \overline{n}$ does not intersect $T^c(\sigma_0)$ transversely, i.e.

$$\overline{n} \geq n^*(\sigma_0) \tag{3.35}$$

By inspecting the phase portrait Fig. 1, a set of fully subsonic solutions can be set up as for (ISP). For obvious reasons we again initialize

$$n = \overline{n}, \quad E = 0 \quad \text{for} \quad \beta = 0 \tag{3.36}$$

The subsonic solution set consists of the trajectory segments starting at $(\overline{n}, E)$ and ending at $(\overline{n}, -E)$, for $E_1 < E < 0$, where $(\overline{n}, E_1)$ is the point of intersection of $T_c(\sigma_0)$ with the line $n = \overline{n}$ in the lower half plane. Then the solution represented by the union of the lower segment of $T_c^{sub}(\sigma_0)$ connecting $(\overline{n}, E_1)$ with $(n_c(\sigma_0), 0)$ and the upper segment connecting $(n_c(\sigma_0), 0)$ with $(\overline{n}, -E_1)$ is adjoined to the solution set.

All these solutions have constant entropy, i.e. (3.33) holds on their respective interval of existence $0 < x < \beta$, where $\beta = \beta(E)$ is the length of travel from $(\overline{n}, E)$ to $(\overline{n}, -E)$. The subsonic subcontinuum constructed above will now be extended by transonic solutions. The situation becomes more involved. Here we sketch the idea of the construction and refer to [8] for details of the proof. After a jump (3.33) does not hold anymore and the right-sided limit of the temperature T_r at the shock is prescribed by the jump condition (3.16),(3.17). In the maximal interval of regularity after the shock, say $[x_0, x_1)$, we have

$$T = \frac{T_r}{n_r^{\frac{2}{3}}} n^{\frac{2}{3}} \tag{3.37}$$

Correspondingly, a new value

$$\sigma_1 = \frac{5}{3J^2} \frac{T_r}{n_r^{\frac{2}{3}}} \tag{3.38}$$

is defined. Then the solution (n, E) is represented by a trajectory segment of (3.25),(3.26) with σ replaced by σ_1.

To make the construction more transparent, we introduce a parameterisation $(n, E) = (n_l(s), \tilde{E}(s))$ of $T_c^{sup}(\sigma_0)$; chosen such that the parameter value $s = 0$ is assumed in the point $(n_c(\sigma_0), 0)$, $s = 1$ in $(n(\sigma_0), 0)$ (where $n_*(\sigma_0)$ denotes the minimal n-value of the trajectory $T_c^{sup}(\sigma_0)$ and $s = 2$ in $(n_c(\sigma_0), 0)$ (at the end of the loop).
We denote $P_c(s) := (n_l(s)\tilde{E}(s))$.

The transonic solutions start out at $(\bar{n}, E_1)$ and follow $T_c(\sigma_0)$ to a point $P_c(s), s > 0$, in the σ_0-supersonic domain. There a jump in (n, T) occurs and the conjugate state $(n_r(s), T_r(s))$ is computed from the jump condition. The state $P^c(s) := (n_r(s), \tilde{E}(s))$ (after the shock) is on the curve $T^c(\sigma_0)$ since E is continuous. In the sequel we shall denote the segment of $T_c(\sigma_0)$, which connects the points $(\bar{n}, E_1)$ and $P^c(s)$ by $T_c(\sigma_0, s)$.

In order to extend the solution for a fixed $s \in (0,2)$ beyond the shock, we have to inspect the $\sigma_1(s)$-phase plane, which is structurally similar to the σ_0-phase plane (it can be shown the $\sigma_0 < \sigma_1(s) < 1 \quad \forall s \in (0,2)$). Now let $s \in (0,2]$ i.e., $\tilde{E}(s) \geq 0$. Then there is a trajectory segment in the $\sigma_1(s)$-phase portrait, which starts at $P^c(s)$ and intersects the line $n = \bar{n}$. This construction of the subcontinuum of transonic single-shock solutions (parameterized in s) can be extended to $s > 1$ (left state $P_c(s)$ with $\tilde{E}(s) < 0$) as long as $P^c(s) \in T^c(\sigma_0)$ is above the branch of $T_c(\sigma_1(s))$ in the lower half plane. It is possible to prove that this is always the case if $\left(\frac{2}{5}\right)^{8/3} \leq \sigma_0 < 1$. The construction of single shock transonic solutions can be continued for $1 \leq s < 2$. For $s = 2$ the smooth transonic solution consisting of the segment of $T_c(\sigma_1(s))$ starting at $(\bar{n}, E_1)$, crossing over into the σ_0-supersonic domain, completing the loop $T_c^{sup}(\sigma_1(s))$ before jumping back into the σ_0-subsonic domain and ending at $(\bar{n}, -E_1)$, has to be joined to the solution set.

By performing sufficiently many loops along $T_c^{sup}(\sigma_1(s))$ before jumping back to the subsonic point P^c, transonic solutions for arbitrarily large values of β can be constructed. All of them except countably many have exactly one shock.

The construction is more complicated in the case $0 < \sigma_0 < \left(\frac{2}{5}\right)^{8/3}$. It can be proven that there exists one and only one $s^* \in (1,2)$ such that the right state $P^c(s^*)$ lies <u>on</u> $T_c^{sup}(\sigma_1(s^*))$. moreover, $P^c(s)$ lies above $T_c^{sup}(\sigma_1(s))$ iff $1 \leq s < s^*$. The construction of single shock transonic solutions described above can be carried out for $1 \leq s < s^*$. For $s = s^*$ we adjoin the following solution to the continuum. It starts out at $(\bar{n}, E_1)$, follows $T_c(\sigma_0)$ into the σ_0- supersonic domain as far as $P_c(s^*)$, jumps to $P^c(s^*) \in T_c^{sub}(\sigma_1(s^*))$, follows the lower branch of $T_c^{sub}(\sigma_1(s^*))$ into $(n_c(\sigma_1(s^*)), 0)$ and the upper branch of $T_c^{sub}(\sigma_1(s^*))$ into its point of intersection with the line $n = \bar{n}$.

The construction proceeds with two-shock transonic solutions. They agree with the solutions constructed above for $s = s^*$ up to the x-value where $T_c^{sub}(\sigma_1(s^*))$ hits $(n_c(\sigma_1(s^*)), 0)$ then, however, they follow $T_c^{sup}(\sigma_1(s^*))$ to a point $(n_l^{(1)}(t), \tilde{E}^{(1)}(t))$ (here $n = n_l^{(1)}(t), E = \tilde{E}^{(1)}(t), 0 < t \leq 2$ is a parameterization of $T_c^{sub}(\sigma_1(s^*))$), and jump to the corresponding conjugate state $(n_l^{(1)}(t), \tilde{E}^{(1)}(t)) \in T^c(\sigma_1(s^*))$. It can be shown that the new σ-value $\sigma_2(t)$ lies in the interval $[(\frac{2}{5})^{8/3}, 1)$ for all $t \in (0,2)$. Thus, the reasoning of Case 1 applies and for every $t \in (0,2)$, there is a trajectory segment in the $\sigma_2(t)$-phase plane, which connects $(n_l^{(1)}(t), \tilde{E}^{(1)}(t))$ and the line $n = \bar{n}$. For $t = 2$ we adjoin the single-shock transonic solution consisting of $T_c(\sigma_0, s^*)$ and the transonic segment of $T_c(\sigma_1, s^*)$ starting at $P^c(s^*)$, performing a full supersonic loop and at the upper point of intersection of $T_c(\sigma_1, S^*)$ with the line $n = \overline{n}$. Fig. 6 shows a typical density n with two shocks and Fig.7 the corresponding two-shocks temperature T. The boundary values $\overline{n} = 3.5$ and $T_0 = 0.11$, and the current density $J = 1$ were chosen.

Obviously, solutions for arbitrarily large values of β can be constructed by performing sufficiently many loops along $T_c^{sup}(\sigma_1, s^*)$ before jumping.

Now denote by (n_e, T_e, E_e) a solution (n, T, E) of (NISP) (for the given value of β) extended to the x-interval $(0, \infty)$ by setting

$$(n_e(x), T_e(x), E_e(x)) = \begin{cases} (n(x), T(x), E(x)), & 0 < x < \beta \\ (0, 0, E(\beta)), & x > \beta \end{cases} \tag{3.39}$$

Simple continuity arguments show that β varies continuously along the above constructed solution set and that the set of extended solutions (n_e, T_e, E_e, β) forms a continuum in the $L^q(0,\infty)^2 \times C[0,\infty] \times R_0^+$-topology for every $1 \leq < \infty$.
We collect the results in:

THEOREM 3.2
Let the assumptions $\sigma_0 < 1$ and $\overline{n} \geq n^*(\sigma_0)$ hold. Then there exists a set S of solutions (n, T, E, β) of (NISP), containing a solution $(n.T, E)$ for every $\beta \geq 0$, such that the set S_e of extended solutions (n_e, T_e, E_e, β) forms a continuous curve in the $L^q(0,\infty)^2 \times C[0,\infty] \times R_0^+$-topology for every $1 \leq \varphi < \infty$. This curve of solutions consists of a bounded subset of subsonic solutions and of an unbounded subcontinuum of transonic solutions. If $\left(\frac{2}{5}\right)^{8/3} \leq \sigma_0 < 1$ holds, then countably many transonic solutions in s are smooth for $x \in [0, \beta]$ and all others have exactly one shock in $(0, \beta]$. If $0 < \sigma_0 < \left(\frac{2}{5}\right)^{8/3}$, then there is a bounded subcontinuum and an unbounded countable subset of single-shock transonic solutions. All other transonic solutions have exactly two shocks in the x-interval $(0, \beta]$.

As in the isentropic model, the case $\overline{n} < n^*(\sigma_0)$ gives rise to solutions with more complicated structures. In particular, the continuum includes solutions with boundary jumps and (depending on the value of σ_0) solutions with one shock and one boundary jump. We refer to [23] for a complete description of this case.

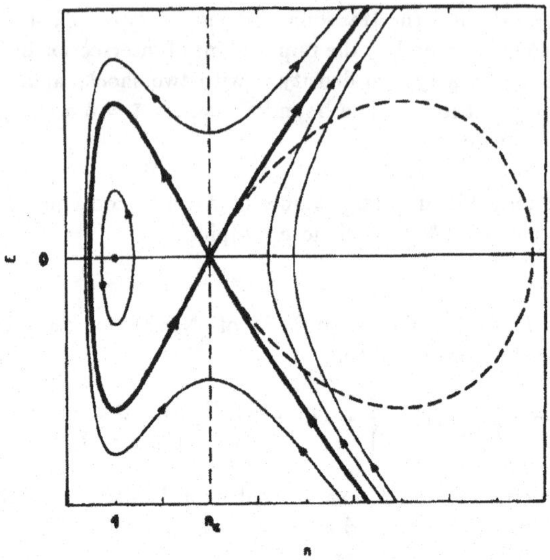

Fig.1 : (n, E) phase-plane for $0 < \sigma < 1$.

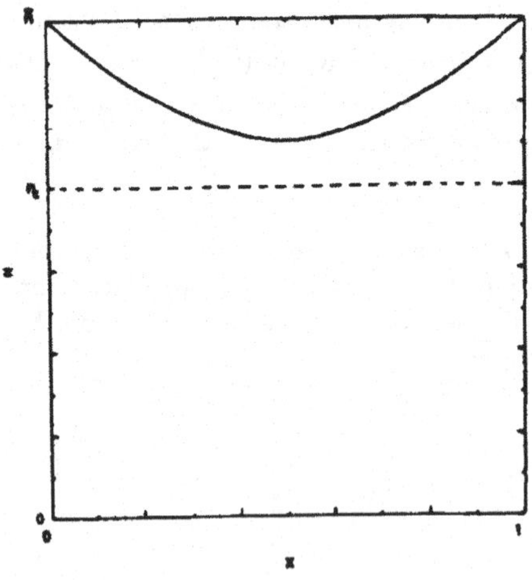

Fig.2 : subsonic solution.

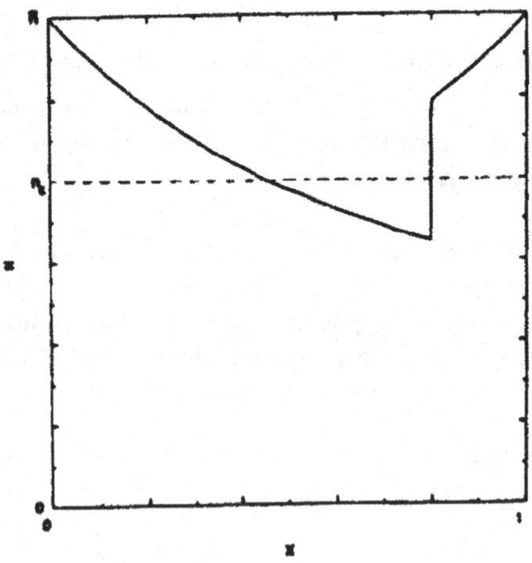

Fig.3 : transonic solution ($E > 0$ at the shock).

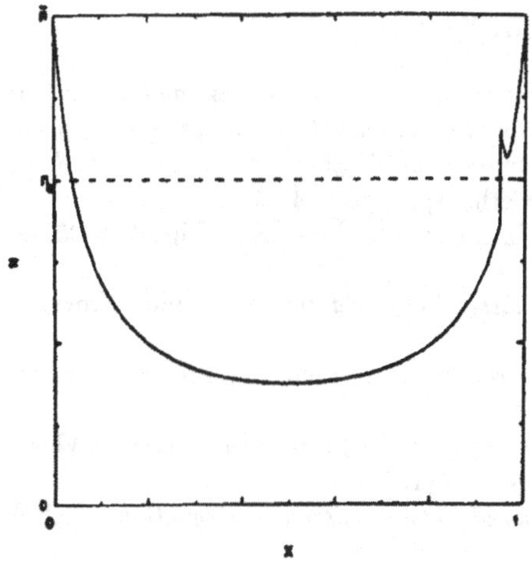

Fig.4 : transonic solution ($E < 0$ at the shock).

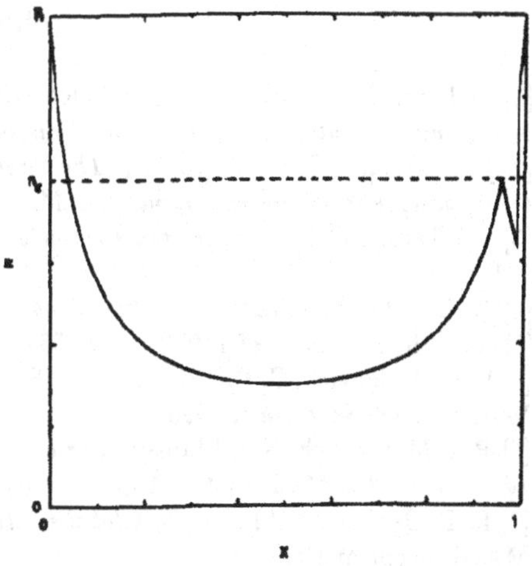

Fig.5 : transonic solution (1 loop along T_e^{sup}).

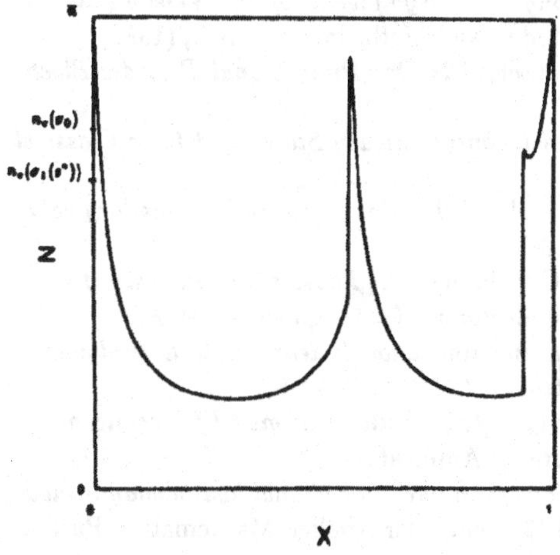

Fig.6 : two-shock transonic electron density $n(x)$.

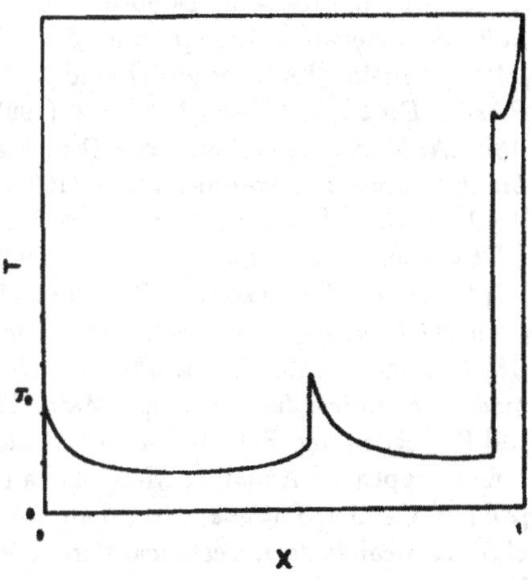

Fig.7 : two shock transonic temperature $T(x)$.

References

[1] R. Illner, H. Lange and P.F. Zweifel, *Global existence and uniqueness and asymptotic behaviour of solutions of the Wigner-Poisson and Schrödinger-Poisson systems*, preprint.

[2] F. Brezzi and P.A. Markowich, *The three-dimensional Wigner-Poisson problem: Existence, uniqueness and approximation*, Math Meth. Appl. Sci. **14** 35-62

[3] V.I. Tatarskii, *The Wigner representation of quantum mechanics*, Sov. Phys.Usp.**26**(1983), 311-327

[4] K. Takahashii, *Distribution functions in classical and quantum mechanics*, Progr. of Theor. Phys. Suppl. **98** (1989), 109-156

[5] P.A. Markowich, C. Ringhofer and C. Schmeiser, "Semiconductor Equations ", Springer Verlag, Wien-New York, 1990

[6] P.A. Markowich, N.J. Mauser, *The classical limit of a self-consistent Quantum-Vlasov equation in 3-d*, Math Meth. Appl. Sci. (to appear) (1992)

[7] R. Di Perna and P.L. Lions, *Global solutions of Vlasov-Poisson type equations*, CEREMADE preprint Nr. 8824.

[8] J. Simon, *Compact sets in the spaces Lp((0,T);B)*, Anal. Math Pura. Appl. **166** (1987), 65-97.

[9] P.L Lions and T. Paul, *Sur les mesures de Wigner*, preprint, CREMADE, Universite de Paris-Dauphine, (1992).

[10] F. Nier, *A stationary Schrödinger-Poisson System arising from the Modelling of Electronic Devices*, Forum Math. **25**, 489-510, (1991).

[11] F. Nier, *A variational Formulation of Schrödinger-Poisson Systems in Dimension* $d \leq 3$, submitted, (1991).

[12] J. Dolbeault, *Stationary States in Plasma Physics; Maxwellian Solutions of the Vlasov-Poisson System*, M^3AS, **1**, 183-208, (1991).

[13] L. Desvillettes & J. Dobbeault *On long time asymptotics of the Vlasov-Poisson-Boltzmann Equation*, to appear in Math. Models and Meth. in Appl. Sci., (1992)

[14] A. Arnold, P.A. Markowich and N. Mauser, *The One-Dimensional Periodic Bloch-Poisson Equation*, M^3AS, **1**, 83-112, (1991).

[15] P.A. Markowich, *Boltzmann Distributed Quantum Steady States and their Classical Limit*, to appear in Forum Math. , (1991).

[16] U. Ascher, J. Christiansen and R.D. Russell, *Collocation software for boundary value ODEs*, Trans. Math. Soft. **7** (1981), 209-222.

[17] U. Ascher, P Markowich, P.Pietra and C. Schmieser, *A Phase plane analysis of transonic solutions for the hydrodynamic semiconductor model*, to appear in M^3AS

[18] P. Degond and P.A. Markowich, *On a one-dimensional steady-state hydrodynamic model for semiconductors*, Appl. Math. Letters **3 (3)**, (1990), 25-86.

[19] P. Degond and P.A. Markowich, *A steady state potential flow model for semiconductors*, to appear in Annali di Matematica Pura ed Applicata.

[20] I.M. Gamba, *Stationary transonic solutions for a one-dimensional hydrodynamic model for semiconductors*, Technical Report **#143**, Center for Applied Mathematics, Purdue University.

[21] I.M. Gamba, *Boundary layer formation for viscosity approximations in transonic flow*, Technical Report **#149**, Center for Applied Mathematics, Purdue University.

[22] P.A. Markowich, "The Stationary Semiconductor Device Equations", Springer-Verlag, Heidelberg, (1986).

[23] P.A. Markowich and P. Pietra, *A non-isentropic Euler-Poisson model for a collisionless plasma*, to appear in Math. Meth. Appl. Sci. , 1993.

[24] P.A. Markowich, C. Ringhofer and C. Schmeiser, "Semiconductor Equations ", Springer

Verlag, Wien-New York, (1990).
[25] J. Smoller, "Shock Waves and Reaction - Diffusion Equations",Springer-Verlag, New York, Heidelberg, Berlin, (1980).

Entropy Methods in Hydrodynamic Scaling *

S. R. S. Varadhan †

1 Introduction. Some easy examples.

The problem of transition from a microscopic description of a system with a large number of components to the description of a suitably scaled version of the same system by a few macroscopic parameters arises often and is physically natural. However the mathematical justification of such a simplified description is considerably more difficult. The purpose of these lectures is to consider several such models and develop some techniques that can be used for a rigorous study of these models.

Our main tool will be entropy and its production. Since entropy production is associated with the existence of stochastic noise in the system we will consider mainly systems with noisy evolution.

The simplest example is the free or noninteracting case. The scaling limit or the passage from a microscopic to a macroscopic description is essentially a law of large numbers. To be precise, consider a one dimensional periodic lattice of N points with a lattice width of $\frac{1}{N}$. We shall assume the lattice points to be arranged uniformly on the boundary of a circle of unit circumference. We have a certain number M of particles that occupy these points or sites. Each particle undergoes a random walk in continuous time on the lattice. With diffusive scaling of space and time we shall assume that each particle waits for an exponential time with expectation $\frac{1}{N^2}$ and then jumps to the site on the left or right (clockwise or counterclockwise) with probability $\frac{1}{2}$. This is done by each particle independently of the other particles. We thus have a system of M particles undergoing independent random walks.

The state of the system at any given time t can be described completely by the position $x_1(t), \ldots, x_M(t)$ of our particles at time t. However if we do not care about the labeling of the particles, the system can be described by the numbers $\{\eta_i(t)\}$ of the particles that occupy the various sites $i = 1, 2, \ldots, N$. If $N \to \infty$ and $M \to \infty$ in such a manner that $\frac{M}{N} \to \bar{\rho}$, then one can think asymptotically of the distribution of total mass $\bar{\rho}$ on the circle S. If $x_1, \ldots, x_M$ is the configuration of the particles and $\eta_1, \ldots, \eta_N$ are the corresponding occupation numbers for the various sites we can for a smooth test function $J(y)$ on S, consider

$$\frac{1}{N}\sum_{i=1}^{M} J(x_i) = \frac{1}{N}\sum J(\frac{i}{N})\eta_i$$

*Work partially supported by NSF grant DMS-9201222.

†Courant Institute, New York University

and if as $N \to \infty$ we have a limit

$$\frac{1}{N} \sum J(\frac{i}{N}) \eta_i \longrightarrow \int_S J(\theta)\, \rho_0(\theta)\, d\theta \tag{1.1}$$

then we can think of $\rho_0(\theta)$ as the asymptotic density of particles. The function $\rho_0(\theta)$ provides the "macroscopic" description of the system and $\{\eta_i\}$ provide the microscopic description. Very many microscopic states lead in the limit to the same macroscopic state.

The problem then is the following: if we know initially that our starting configuration of particles led to a macroscopic density $\rho_0(\theta)$, can we conclude anything about the macroscopic state at a later time? If the macroscopic state is $\rho(t, \theta)$ at time t, how do we compute $\rho(t, \theta)$ from $\rho_0(\theta)$? Underlying the question is the unsupported assumption that if two different microscopic states lead to the same macroscopic state at time $t = 0$ then they lead to the same macroscopic state at times $t > 0$, so that the macroscopic state at $t > 0$ is determined if we know only the initial macroscopic state.

It is the underlying assumption that makes the task harder. However in our current model of noninteracting particles everything is quite elementary and can be seen by a straightforward easy computation. Let $p^{(N)}(t, \frac{i}{N}, \frac{j}{N})$ be the transition probability of a single particle from site $\frac{i}{N}$ to site $\frac{j}{N}$ during time t. The central limit theorem tells us that

$$\lim_{N\to\infty} \sum_{j=1}^{N} p^{(N)}(t, \frac{i}{N}, \frac{j}{N})\, F(\frac{j}{N}) \to \int_S p(t, x, y)\, F(y)\, dr \tag{1.2}$$

provided $\frac{i}{N} \to x$ as $N \to \infty$ and F is nice. Let us take a test function G on S and consider the random variable

$$\frac{1}{N} \sum_{k=1}^{N} G(x_k(t)) = \frac{1}{N} \sum G(\frac{i}{N})\, \eta_i(t) = \xi_N(t)$$

Let us think of $x_1(0), \ldots, x_M(0)$ as given and nonrandom. Otherwise we have to condition with respect to them.

$$E\{\xi_N(t)\} = \frac{1}{N} \sum_{k=1}^{N} \left\{ \sum_{j=1}^{N} G(\frac{j}{N})\, p(t, x_k(0), \frac{j}{N}) \right\}$$

and $\lim_{N\to\infty} E\{\xi_N(t)\}$ exists and equals

$$\lim_{N\to\infty} E\{\xi_N(t)\} = \int \left(\int G(y)\, p(t, x, y)\, dy \right) \rho_0(x)\, dx \tag{1.3}$$

where $\rho_0(\theta)$ is given by (1.1). If we define

$$\rho(t, y) = \int p(t, x, y)\, \rho_0(x)\, dx \tag{1.4}$$

then

$$\lim_{N\to\infty} E\{\xi_N(t)\} = \int G(y)\, \rho(t, y)\, dy \tag{1.5}$$

We can also calculate

(1.6)

$$
\begin{aligned}
\lim_{N\to\infty} E[\xi_N(t)]^2 &= \lim_{N\to\infty} E\left\{\tfrac{1}{N^2}\sum G(x_k(t))\,G(x_k(t))\right\}\\
&= \lim_{N\to\infty} \tfrac{1}{N^2}\sum_{k\neq \ell} E\left\{G(x_k(t)))\,G(x_\ell(t))\right\}\\
&= \lim_{N\to\infty} \tfrac{1}{N^2}\sum_{k\neq \ell}\left[\textstyle\int G(y)\,p(t,x_k(0),y)\,dy \int G(y)\,p(t,x_\ell(0),y)\,dy\right]\\
&= \lim_{N\to\infty}\left(\tfrac{1}{N}\sum \textstyle\int G(y)\,p(t,x_k(0),y)\,dy\right)^2\\
&= \left(\textstyle\int\left(\int G(y)\,p(t,x,y)\,dy\right)\rho_0(x)\,dx\right)^2\\
&= \left(\textstyle\int G(y)\,\rho(t,y)\,dy\right)^2
\end{aligned}
$$

(1.5) and (1.6) yield the law of large numbers

$$\lim_{N\to\infty} \xi_N(t) = \int G(y)\,\rho(t,y)\,dy$$

in probability. This tells us that $\rho(t,y)$ is the macroscopic state at time t and can be obtained from $\rho_0(y)$ by the formula

$$\rho(t,y) = \int p(t,x,y)\,\rho_0(x)\,dx \tag{1.7}$$

or equivalently by solving the heat equation

$$\frac{\partial\rho}{\partial t} = \frac{1}{2}\rho_{yy} \quad \text{for} \quad t>0\,, \quad y\in S \tag{1.8}$$

with initial condition $\rho(t,y)|_{t=0} = \rho_0(y)$.

A second example which has some interaction is the symmetric random walk with exclusion. The model is similar to the earlier one except it is not allowed to have more than one particle at any given site. If a particle decides to jump to a new site according to the earlier description, i.e. waiting for an exponential time and then picking one of the two neighboring sites with equal probability, and if the site is already occupied then the particle cannot jump and must wait all over again for another exponential duration.

If we denote by $\eta_i(t)$, $i = 1,2,\ldots,N$, the occupancy numbers which are either 0 or 1 according to the site i being empty or occupied at time t, it is easy to write down the infinitesimal generator of our process in the "η" variables.

$$(\mathcal{A}f)(\eta) = \frac{N^2}{2}\sum[\eta_i(1-\eta_{i+1})+\eta_{i+1}(1-\eta_i)]\left[f(\eta^{i,i+1})-f(\eta)\right] \tag{1.9}$$

where $\eta^{i,i+1}$ is the new configuration defined by

$$
\begin{aligned}
\eta_j^{i,i+1} &= \eta_j &&\text{if} \quad j\neq i \text{ or } i+1\\
&= \eta_{i+1} &&\text{if} \quad j=i\\
&= \eta_i &&\text{if} \quad j=i+1
\end{aligned}
$$

It turns out that

$$
\begin{aligned}
\eta_i(1-\eta_{i+1})+\eta_{i+1}(1-\eta_i) &= 1 \quad \text{if} \quad \eta_i\neq\eta_{i+1}\\
&= 0 \quad \text{if} \quad \eta_i=\eta_{i+1}
\end{aligned}
$$

and

$$f(\eta^{i,i+1}) = f(\eta) \quad \text{if} \quad \eta_i = \eta_{i+1}$$

for any $f(\cdot)$ because the configuration $\eta^{i,i+1} \equiv \eta$ in such a case. Therefore one can rewrite $\mathcal{A}$ in the more convenient form

$$(1.10) \qquad (\mathcal{A}f)(\eta) = \frac{N^2}{2} \sum \left[(\eta^{i,i+1}) - f(\eta)\right]$$

Let us define the sum

$$(1.11) \qquad z(t) = \frac{1}{N} \sum J(\frac{i}{N})\, \eta_i(t) = \psi(\eta(t))$$

where $\psi(\eta) = \frac{1}{N} \sum J(\frac{i}{N})\eta_i$ for some smooth test function $J(\cdot)$ on S. We can calculate

$$(1.12) \qquad dz(t) = (\mathcal{A}\psi)(\eta(t))\, dt + d\xi_N(t)$$

where

$$(1.13) \qquad \begin{aligned} (\mathcal{A}\psi)(\eta) &= \tfrac{N^2}{2} \cdot \tfrac{1}{N} \sum \left[J(\tfrac{i+1}{N}) - J(\tfrac{i}{N})\right] [\eta_{i+1} - \eta_i] \\ &= \tfrac{1}{N} \cdot \tfrac{N^2}{2} \cdot \sum \left[J(\tfrac{i+1}{N}) - 2J(\tfrac{i}{N}) + J(\tfrac{i-1}{N})\right] \eta_i \\ &\simeq \tfrac{1}{2N} \sum J''(\tfrac{i}{N})\eta_i \end{aligned}$$

and $d\xi_N(t)$ is a martingale term. An elementary computation yields that for any finite T

$$(1.14) \qquad E|\xi_N(T) - \xi_N(0)|^2 \to 0 \quad \text{as} \quad N \to \infty$$

irrespective of the initial configuration $\{\eta_i(0);\ 1 \le i \le N\}$.

One can therefore conclude immediately that

$$(1.15) \quad \lim_{N\to\infty} \left[\frac{1}{N} \sum J(\frac{i}{N})\, \eta_i(T) - \frac{1}{N} \sum J(\frac{i}{N})\, \eta_i(0) - \frac{1}{2N} \int_0^T \sum J''(\frac{i}{N})\, \eta_i(s)\, ds\right] = 0$$

in probability. It is routine to establish from this point that any limit point in the sense of weak convergence of the distribution of the measure valued stochastic process $\frac{1}{N} \sum \delta_{i/N}\eta_i(t)$, on $[0,T]$ will be concentrated on the set of densities $\rho(\cdot, y)$ on $[0,T]$ that satisfy in a weak sense the heat equation

$$\frac{\partial \rho}{\partial t} = \frac{1}{2} \frac{\partial^2 \rho}{\partial y^2}, \qquad 0 \le t \le T$$

compatible with the initial data $\rho(0, y) = \rho_0(y)$ that is defined as the limit in probability of

$$\frac{1}{N} \sum \delta_{i/N}\eta_i(0)$$

in the sense of weak convergence. From the uniqueness of solutions to the heat equation we obtain the fact that

$$\frac{1}{N} \sum \delta_{i/N}\eta_i(t)$$

is close to $\rho(t, y)\, dy$ in the sense of weak convergence (in probability). This is easily made formal by a combination of steps that we will outline once for the record.

Consider $\frac{1}{N}\sum \delta_{i/N}\eta_i(t) = \lambda_N(t,\omega,dy)$ as a function on $[0,T]$ with values in the space $\mathcal{M}(S)$ of probability measures on S. Since the paths are right continuous, we have corresponding to our basic stochastic process P_N on the configuration space $\{\eta_i(t)\ 0 \le t \le T,\ 1 \le i \le N\}$ a stochastic process Q_N on the space $D[[0,T],\mathcal{M}(S)]$ of measure valued processes with discontinuities of the first kind only. We first establish the compactness of the family $\{Q_N\}$. The total mass

$$\frac{1}{N}\sum \eta_i(t) = \lambda_N(t,\omega,S) \le 1 \quad \text{for all} \quad t \ge 0 \text{ and } \omega\ .$$

Lemma 1.1. *For any function $J(y)$ that is smooth on S and $\epsilon > 0$*

$$\limsup_{\delta\to 0}\ \limsup_{N\to\infty}\ P_N\left[\sup_{\substack{0\le s<t\le T\\ |t-s|<\delta}} \left|\frac{1}{N}\sum J(\frac{i}{N})\,\eta_i(t) - \frac{1}{N}\sum J(\frac{i}{N})\,\eta_i(s)\right| \ge \epsilon\right] = 0\ .$$

Proof: By the properties of the generator

$$\begin{aligned}&\tfrac{1}{N}\sum J(\tfrac{i}{N})\,\eta_i(t) - \tfrac{1}{N}\sum J(\tfrac{i}{N})\,\eta_i(s)\\ &\quad = \tfrac{1}{2N}\int_s^t \sum N^2\left[J(\tfrac{i-1}{N}) - 2J(\tfrac{i}{N}) + J(\tfrac{i+1}{N})\right]\eta_i(\sigma)\,d\sigma\\ &\quad +\xi_J^{(N)}(t) - \xi_J^{(N)}(s)\ .\end{aligned}$$

We assume without loss of generality that $\xi_J^{(N)}(0) = 0$.

$$\left|\frac{1}{2N}N^2\sum\left[J(\frac{i-1}{N}) - 2J(\frac{i}{N}) + J(\frac{i+1}{N})\right]\eta_i(\sigma)\right| \le C$$

where C is a constant depending only on J. Moreover

$$\begin{aligned}P_N\left\{\textstyle\sum_{0\le s\le T}|\xi_J^{(N)}(t)| \ge \epsilon\right\} &\le \tfrac{1}{\epsilon^2}E^{P_N}\left\{|\xi_J^{(N)}(T)|^2\right\}\\ &= o(1) \quad \text{as} \quad N\to\infty\ .\end{aligned}$$

Combining the two estimates we have Lemma 1.1.

The space $D[[0,T],\mathcal{M}(S)]$ has its natural Skorohod topology. Because $\mathcal{M}(S)$ is compact Lemma 1 ensures that Q_N is compact and in fact any limit point Q is supported on $C[[0,T],\mathcal{M}(S)]$.

Lemma 1.2. *Any limit point Q is supported on the set of solutions of the equation*

$$\frac{\partial\rho}{\partial t} = \frac{1}{2}\rho_{xx}\ .$$

Proof: Because the martingale term $\xi_J^{(N)}(t)$ goes to zero as $N\to\infty$, in the limit if $\lambda(t,dx)$ is an element of $C[[0,T],\mathcal{M}(S)]$

$$\frac{\partial}{\partial t}\int J(x)\,\lambda(t,dx) = \frac{1}{2}\int J''(x)\,\lambda(t,dx)\ ,$$

because of the exclusion property in the limit

$$\lambda(t,A) \le |A|$$

so that $\lambda(t,dx) = \rho(t,x)\,dx$ with $0 \le \rho \le 1$. Our measure $\lambda(t,dx)$ is then a weak solution $\rho(t,x)$ of

$$\frac{\partial \rho}{\partial t} = \frac{1}{2}\rho_{xx}\,.$$

Lemma 3. *If at $t = 0$ we have a weak law of large numbers*

$$\lim_{N\to\infty} \frac{1}{N}\sum J(\frac{i}{N})\,\eta_i(0) = \int J(x)\,\rho_0(x)\,dx$$

for some deterministic $\rho_0(x)$ then Q is supported on the set of densities $\rho(t,x)$ that satisfy

$$\rho(0,x) = \rho_0(x) \text{ a.e. } x\,.$$

Proof: Obvious.

Combining the three lemmas with the uniqueness of weak solutions to the heat equation we see that the limiting measure Q can be supported by only one element of $C[[0,T], \mathcal{M}(S)]$ namely $\rho(t,x)\,dx$ where ρ solves the heat equatin with the given initial data.

2 Ginsburg-Landau type model (two lectures).

We shall consider a model which is more complicated than the ones we considered in the previous lecture. We have a variable x_i, $1 \le i \le N$ attached to N successive sites of a one dimensional periodic lattice. We will scale the lattice by a factor of $\frac{1}{N}$ and denote the sites by $\frac{i}{N}$ representing points θ on the circumference of a circle S of unit length. The collection $\{x_1,\ldots,x_N\}$ viewed as a point of R^N will undergo in time a diffusion determined by the following system of stochastic differential equations.

$$(2.1)\qquad \begin{cases} dx_i(t) &= dz_{i-1,i}(t) - dz_{i,i+1}(t) \\ dz_{i,i+1}(t) &= \frac{N^2}{2}\,[c(x_i(t)) - c(x_{i+1}(t))]\,dt + N\,d\beta_{i,i+1}(t) \end{cases}$$

where

$$c(x) = \phi'(x) = \frac{d\phi(x)}{dx}$$

for a suitable potential $\phi(x)$ satisfying properties to be listed later. We can write the equations more concisely in the form

$$(2.2)\qquad dx_i(t) = \frac{N^2}{2}\nabla^2\,[\phi'(x(t))]\,dt + N\,[d\beta_{i-1,i}(t) - d\beta_{i,i+1}(t)]$$

here $\beta_{i,i+1}(t)$ are N independent Brownian motions. The generator for the diffusion is

$$(2.3)\qquad L_N = \frac{N^2}{2}\sum\left(\frac{\partial}{\partial x_i} - \frac{\partial}{\partial x_{i+1}}\right)^2 - \frac{N^2}{2}\sum[\phi'(x_i) - \phi'(x_{i+1})]\left[\frac{\partial}{\partial x_i} - \frac{\partial}{\partial x_{i+1}}\right]$$

The factor N^2 is the effect of speeding up time by N^2, to correspond with the rescaling of space by a factor of N, under diffusive scaling. One should think of $\phi(x) = \frac{x^2}{2}$ as a typical case. We make the following assumptions on $\phi(x)$

$$(2.4)\qquad \int e^{-\phi(x)}\,dx = 1$$

This is only a normalization condition. Since ϕ is only defined up to an additive constant the real assumption is

$$\int e^{-\phi(x)}\,dx < \infty\,. \tag{2.4$'$}$$

$$\int e^{\lambda x-\phi(x)}\,dx = M(\lambda) < \infty \quad \text{for all} \quad \lambda \in R\,. \tag{2.5}$$

$$\int e^{\sigma|\phi'(x)|-\phi(x)}\,dx < \infty \quad \text{for all} \quad \sigma > 0\,. \tag{2.6}$$

We assume that $\phi'(x)$ is continuous. Under the above assumptions the process can be shown to exist as a well defined diffusion. Moreover if we define

$$\begin{aligned} a_n &= \max\left(\int e^{n|x|-\phi(x)}\,dx\,,\ \int e^{n|\phi(x)|-\phi(x)}\,dx\right)\\ \omega(|x|) &= 1+\tfrac{1}{2}\log\textstyle\sum_{n=0}^{\infty} e^{n|x|}\cdot\frac{1}{2^n}\cdot\frac{1}{a_n}\,, \end{aligned}$$

then $\omega(|x|)$ is nondecreasing in $|x|$, symmetric, $\omega(0) \geq 1$ and

$$\lim_{|x|\to\infty} \frac{|x|}{\omega(|x|)} = 0 \tag{2.7}$$

Furthermore if

$$W(x) = \omega(|x|) + \omega(|\phi'(x)|)$$

then

$$e^{W(x)} \leq \frac{1}{2}\left[e^{2\omega(|x|)} + e^{2\omega(|\phi'(x)|)}\right]$$

and

$$\int e^{W(x)-\phi(x)}\,dx \leq 2e^2 = C < \infty \tag{2.8}$$

The process constructed has the following properties:

1) L_N is reversible with respect to the weight $e^{-\Sigma\phi(x_i)}\,dx$ which we shall denote by μ_N. μ_N is of course an invariant probability measure on R^N for our L_N.

2) The quantity $x_1 + \cdots + x_N$ is a constant in time for our diffusion and therefore μ_N is not ergodic for our diffusion.

3) However the conditionals $\mu_{N,y}$ of μ_N on the hyperplanes $\frac{x_1+\cdots+x_N}{N} = y$, well defined for every y are invariant and ergodic for every y. This is a consequence of the ellipticity of L_N on each hyperplane.

4) We have a Dirichlet form corresponding to L_N and it is given by

$$D_N(u) = \frac{N^2}{2}\int \sum\left(\frac{\partial u}{\partial x_i} - \frac{\partial u}{\partial x_{i+1}}\right)^2 d\mu_N \tag{2.9}$$

The function $M(\lambda)$ defined by (1.5) has the property that $\log M(\lambda)$ is convex in λ and we can define the conjugate convex function

$$h(y) = \sup_\lambda\,[\lambda y - \log M(\lambda)] \tag{2.10}$$

By familiar results

$$y = \frac{M'(\lambda)}{M(\lambda)} \quad \text{and} \quad \lambda = h'(y)$$

are inverse functions of each other and are regular.

Suppose we start with an initial distribution for $x_1, \ldots, x_N$ given by the density $f_N^0(x_1, \ldots, x_N)$ with respect to μ_N then the density will evolve over time given by a density $f_N^t(x_1, \ldots, x_N)$ at time t according to the equation

$$\left\{ \begin{array}{l} \frac{\partial f_N^t}{\partial t} = L_N f_N^t \\ f_N^t|_{t=0} = f_N^0 \end{array} \right. \tag{2.11}$$

We assume that at time 0, there is a deterministic profile $m_0(\theta)$, an integrable function on S, such that

$$\frac{1}{N} \sum J(\frac{i}{N}) x_i \to \int J(\theta)\, m_0(\theta)\, d\theta \tag{2.12}$$

in probability relative to $f_N^0\, d\mu_N$ for all continuous test functions on S. We wish to conclude under mild additional restrictions on f_N^0 that

$$\frac{1}{N} \sum J(\frac{i}{N}) x_i \to \int J(\theta)\, m(t, \theta), d\theta \tag{2.13}$$

again for all continuous test functions on S, relative to the distribution $f_N^t\, d\mu$. Moreover we need to specify how $m(t, \theta)$ evolves from $m_0(\theta)$ over time. The aim of this lecture and the next is to prove that we can conclude (1.13) with $m(t, \theta)$ chosen as the unique solution of

$$\frac{\partial m}{\partial t} = \frac{1}{2} (h'(m))_{\theta\theta}, \qquad t > 0, \quad \theta \in S \tag{2.14}$$

$$\text{with} \quad m(t, \theta)|_{t=0} = m_0(\theta) \quad \text{on} \quad S \tag{2.15}$$

The above result is proved in [5] which we will follow. A proof in a modified situation under more restrictive conditions can be found in [3] and [4]. It is proved in [14] by another method to which we will return in Lecture 4. The mild assumption we will make on $f_N^0(x_1, \ldots, x_N)$ is that there exist a finite constant C such that

$$\int f_N^0 \log f_N^0\, d\mu_N \leq CN \quad \text{for all} \quad N \tag{2.16}$$

(The constant C is independent of N.)

An easy computation yields for any $J(\theta)$ on S which is twice continuously differentiable

$$\begin{aligned} d(\tfrac{1}{N} \textstyle\sum J(\tfrac{i}{N})\, x_i(t)) &= \tfrac{N^2}{2} \cdot \tfrac{1}{N} \textstyle\sum J(\tfrac{i}{N})\, [\nabla^2 \phi'(x_i(t))]\, dt \\ &\quad + \textstyle\sum J(\tfrac{i}{N})\, \{d\beta_{i-1,i} - d\beta_{i,i+1}\} \\ &\simeq \tfrac{1}{2N} \textstyle\sum J''(\tfrac{i}{N})\, \phi'(x_i(t))\, dt \\ &\quad + \tfrac{1}{N} \textstyle\sum J' \tfrac{i}{N})\, d\beta_{i,i=1}(t) \\ &= \tfrac{1}{2N} \textstyle\sum J''(\tfrac{i}{N})\, \phi'(x_i(t))\, dt + d\xi_J^{(N)}(t) \end{aligned} \tag{2.17}$$

where

$$\xi_J^{(N)}(t) = \frac{1}{N}\int_0^t \sum J'(\frac{i}{N})\, d\beta_{i,i+1}(t)\,. \tag{2.18}$$

There are some errors in replacing differences of J with derivatives. These are easily controlled.

Let us denote by $\mathcal{M}$ the set of signed measures on S of finite variation. We will give it the weak $*$ topology as the dual of $C(S)$, the space of continuous functions on S. Although $\mathcal{M}$ is not such a good space, the subsets $\mathcal{M}_\ell$ of $\mathcal{M}$ consisting of signed measures of variation at most ℓ is a compact metric space. $\mathcal{M}$ will be treated mainly as the union of $\mathcal{M}_\ell$ over ℓ. Let us fix an arbitrary finite interval $[0,T]$ and consider the space Ω_ℓ of continuous maps of $[0,T]$ into Ω_ℓ and the set $\Omega = \bigcup_\ell \Omega_\ell$. Given a point $(x_1,\ldots,x_N)$ in R^N we can map it into $\mathcal{M}$ by

$$(x_1\,,\,\ldots\,,\,x_N) \to \frac{1}{N}\sum_{i=1}^{N} x_i\delta_{i/N}$$

and the trajectory $(x_1(t),\ldots,x_N(t))$ can be mapped into $\frac{1}{N}\sum x_i(t)\,\delta_{i/N}$ for $0 \le t \le T$. We denote this element of Ω by $\nu(t,d\theta)$, $0 \le t \le T$ or by $\nu(\cdot,d\theta)$. Our basic stochastic process P_N^0 with initial distribution $f_N^0\,d\mu_N$ will be mapped into a measure Q_N^0 on Ω. We can also initially start in equilibrium μ_N and this will give a stationary P_N which is mapped into a stationary process Q_N on Ω.

First we establish the compactness of Q_N^0 on Ω.

Theorem 2.1. *The following hold:*

$$\lim_{\ell\to\infty}\limsup_{N\to\infty} Q_N^0\left[\sup_{0\le t\le T}\|\nu(t,\cdot)\| \ge \ell\right] = 0$$

or equivalently

$$\lim_{\ell\to\infty}\limsup_{N\to\infty} P_N^0\left[\sup_{0\le t\le T}\frac{1}{N}\sum|x_i(t)| \ge \ell\right] = 0 \tag{2.19}$$

$$\lim_{\delta\to 0}\limsup_{N\to\infty} P_N^0\left[\sup_{\substack{0<s<t\le T\\ t-s\le\delta}}\int_s^t \frac{1}{N}\sum|\phi'(x_i(\sigma))|\,d\sigma \ge \epsilon\right] = 0 \tag{2.20}$$

for every $\epsilon > 0$

(2.21) *For every* $\epsilon > 0$ *and for every nice* $J(\cdot)$
$\lim_{N\to\infty} P_N^0\left[\sup_{0\le t\le T}|\frac{1}{N}\int_0^T\sum J'(\frac{i}{N})\,d\beta_{i,i+1}(t)| \ge \epsilon\right] = 0\,.$

Proof of (2.19): This is based on the following estimate from the theory of Dirichlet forms. Let $u(x)$ be a nice function on the state space with finite Dirichlet and L_2-norms with respect to a reversible Markov Process with invariant probability measure μ. If P refers to the stationary process with invariant measure μ, then

$$P\left[\sup_{0\le t\le T}|U(x(t))| \ge \ell\right] \le \frac{e}{\ell}\left(\|u\|^2 + T\|u\|_1^2\right)^{1/2}$$

where $\|u\|_0^2$ and $\|u\|_1^2$ are respectively the L_2 and Dirichlet norms of $u(\cdot)$. We apply this with

$$u(x_1,\dots,x_N) = \exp[|x_1| + \cdots + |x_N|]$$

and $e^{N\ell}$ for ℓ.

An easy computation yields:

$$\begin{aligned} &P_N\left[\sup_{0\le t\le T} \tfrac{1}{N}\textstyle\sum_1^N |x_i(t)| \ge \ell\right] \\ &\le P_N\left[\sup_{0\le t\le T} \exp\left[\tfrac{1}{N}\textstyle\sum |x_i(t)|\right] \ge \exp \ell\right] \\ &= P_N\left[\sup_{0\le t\le T} \exp\left[\textstyle\sum |x_i(t)\right] \ge \exp N\ell\right] \\ &\le e\cdot e^{-N\ell}(C_1^N + N^2 C_2 \epsilon_1^N)^{1/2} \\ &\le C_4\, e^{-N\ell + C_5 N} \end{aligned}$$

By our hypothesis (2.16) on f_N^0, the relative entropy

$$H(P_N^0 \mid P_N) \le CN \tag{2.22}$$

and by the inequality

$$P_N^0(A) \le \frac{H(P_N^0 \mid P_N) + 1}{\log \frac{1}{P_N(A)}}$$

valid for every A, if $\ell \ge C_5$

$$P_N^0(A) \le \frac{CN + 1 + \log C_5}{N\ell - C_5 N} \tag{2.23}$$

we first let $N \to \infty$ and then $\ell \to \infty$ to get (a).

Proof of (2.20): Because of (2.7) for each $\epsilon > 0$, there exists a constant C_ϵ such that

$$\begin{aligned} |\phi'(x)| &\le \epsilon\omega(|\phi'(x)|) + C_\epsilon \\ &\le \epsilon W(x) + C_\epsilon\,. \end{aligned}$$

Therefore for $0 < s < t < T$ with $t - s < \delta$

$$\begin{aligned} \tfrac{1}{N}\textstyle\int_s^t \sum |\phi'(x_i(\sigma))|\, d\sigma &\le C_\epsilon|t-s| + \tfrac{\epsilon}{N}\textstyle\int_s^t \sum |W(x_i(\sigma)|\, d\sigma \\ &\le C_\epsilon|t-s| + \tfrac{\epsilon}{N}\textstyle\int_0^T \sum |W(x_i(\sigma))|\, d\sigma \\ &\le C_\epsilon\delta + \tfrac{\epsilon}{N}\textstyle\int_0^T \sum W(x_i(\sigma))\, d\sigma \end{aligned} \tag{2.24}$$

For the function W

$$\begin{aligned} E^{P_N}\left\{\exp[\tfrac{1}{T}\textstyle\sum \int_0^T W(x_i(\sigma))\, d\sigma\right\} &\le E^{P_N}\left\{\tfrac{1}{t}\textstyle\int_0^T \cdot \sum \exp\left[W(x_i(\sigma))\right]\, d\sigma\right\} \\ &= \textstyle\int \exp \sum W(x_i)\cdot d\mu_N \\ &= C^N \end{aligned}$$

By the entropy inequality

$$E^{P_N^0}(F) \le \log E^{P_N}[e^F] + H(P_N \mid P_N^0)$$

applied to

$$F = \frac{1}{T}\int_0^T \sum W(x_i(\sigma))\, d\sigma$$

$$E^{P_N^0} \int_0^T \sum W(x_i(\sigma))\, d\sigma \leq TN(C + \log C) \leq C_T \cdot N \tag{2.25}$$

Combining it with (2.24)

$$\begin{aligned} &E^{P_N^0}\left[\sup_{\substack{0 \leq s < t \leq T \\ t-s \leq \delta}} \int_s^t \frac{1}{N} \sum |\phi'(x_i(\sigma))|\, d\sigma\right] \\ &\leq E^{P_N^0}\left[C_\epsilon \delta + \frac{\epsilon}{N} \int_0^T \sum W(x_i(\sigma))\, d\sigma\right] \\ &\leq C_\epsilon \delta + \epsilon C_T\,. \end{aligned} \tag{2.26}$$

We let $\delta > 0$. Since $\epsilon > 0$ is arbitrary we are done.

(2.21) is an easy consequence of Doob's inequality for martingales and the computation

$$E^{P_N^0} |\xi_J^N(T)|^2 \leq \frac{C_J}{N}\,.$$

Having established the compactness of $\{Q_N^0\}$ we can take a weak limit and call it Q. We will prove that Q has the following properties.

(A) The measure $\nu(t, d\theta)$ has a density $m(t, \theta)$ for almost all $\nu(\cdot, d\theta)$ with respect to Q. In fact the exceptional set does not depend on $t \in [0, T]$.

(B) The measure Q is supported on those densities $m(t, \theta)$ that satisfy

$$m(0, \theta) = m_0(\theta) \text{ a.e. } \theta$$

$$\int J(\theta) m(t, \theta)\, d\theta - \int J(\theta) m_0(\theta)\, d\theta = \frac{1}{2} \int \int_0^t J''(\theta)\, h'(m(s, \theta))\, d\theta\, ds$$

for all twice continuously differentiable test functions $J(\cdot)$ on S.

(C)

$$E^Q \int_0^T \int \left[\frac{\partial}{\partial \theta} h'(m(t, \theta))\right]^2 d\theta\, dt = C < \infty\,.$$

From known uniqueness theorems for solutions of

$$\frac{\partial m}{\partial t} = \frac{1}{2} (h'(m))_{\theta\theta}$$

in the class of solutions satisfying

$$\int_0^T \int \left[\frac{\partial}{\partial \theta} h'(m(t, \theta))\right]^2 dt\, d\theta < \infty$$

the hydrodynamic scaling limit will follow and we will have the theorem

Theorem 2.2. *The limit measure Q is concentrated at the unique trajectory in Ω solving*

$$\frac{\partial m}{\partial t} = \frac{1}{2} (h'(m))_{\theta\theta}$$

with initial condition $m_0(\theta)$ so that for any continuous test function $J(\cdot)$ on S

$$\begin{aligned}&\lim_{N\to\infty} Q_N^0 \left[\sup_{0\le t\le T} |\textstyle\int J(\theta)\nu_i(t,d\theta) - \int J(\theta)m(t,\theta)\,d\theta| \ge \epsilon\right]\\ &= \lim_{N\to\infty} P_N^0 \left[\sup_{0\le t\le T} \left|\frac{1}{N}\sum J(\frac{i}{N})\,x_i(t) - \int J(\theta)m(t,\theta)\,d\theta\right| \ge \epsilon\right]\\ &= 0\,.\end{aligned}$$

The rest of the lecture will be devoted to a proof of (A), (B) and (C).

Proof of (A): We start with the relation

$$E^{P_N}\left\{\exp\sum J(\frac{i}{N})\,x_i(t)\right\} = \exp\left[\sum \log M(J(\frac{i}{N}))\right]$$

from which we deduce

$$\lim_{N\to\infty}\frac{1}{N}\log E^{P_N}\left\{\exp\left[\sum J(\frac{i}{N})\,x_i(t)\right]\right\} \le \int \log M(J(\theta))\,d\theta \tag{2.27}$$

We use entropy to inequality to conclude

$$E^{P_N^0}\left\{\frac{1}{N}\sum J(\frac{i}{N})\,x_i(t)\right\} \le \frac{1}{N}\sum \log M(J(\frac{i}{N})) + C \tag{2.28}$$

or

$$E^{Q_N^0}\left\{\int J(\theta)\,\nu(t,d\theta) - \frac{1}{N}\sum \log M(J(\frac{i}{N})\right\} \le C$$

Passing to the limit as $N\to\infty$

$$E^{Q}\left\{\int J(\theta)\,\nu(t,d\theta) - \int \log M(J(\theta))\,d\theta\right\} \le C\,. \tag{2.29}$$

To pass to the limit from (2.27) to (2.28) we need the uniform integrability of

$$\frac{1}{N}\sum J(\frac{i}{N})\,x_i(t)$$

with respect to P_N^0. Because of (2.7) it is sufficient to prove the boundedness of

$$E^{P_N^0}\left[\frac{1}{N}\sum W(x_i(t))\right]$$

or by entropy inequality, establish the boundedness of

$$\begin{aligned}&\frac{1}{N}\log E^{P_N}\left[\exp\left\{\textstyle\sum W(x_i(t))\right\}\right]\\ &= \frac{1}{N}\log \textstyle\int \exp\left[\sum W(x_i)\right]\,d\mu_N\\ &= \log \textstyle\int \exp\left[W(x) - \phi(x)\right]\,dx\\ &= C < \infty\,.\end{aligned}$$

The estimate (2.28) is valid for every $J(\cdot)$ and therefore

$$\sup_{J(\cdot),t} E^{Q}\left\{\int J(\theta)\,\nu(t,d\theta) - \int \log M(J(\theta))\,d\theta\right\} \le C \tag{2.30}$$

If we can move the supremum inside we will have

$$(2.31) \qquad E^Q\left\{\sup_{J(\cdot),t}\int J(\theta)\,\nu(t,d\theta) - \int \log M(J(\theta))\,d\theta\right\} \le C$$

But this is the same as

$$(2.32) \qquad E^Q\left\{\sup_{0\le t\le T}\int h(m(t,\theta))\,d\theta\right\} \le C$$

thereby implicitly proving by duality that $\nu(t,d\theta) \ll d\theta$ with density $m(t,\theta)$, satisfying (2.31). We shall therefore concentrate on moving the supremum inside from (2.29) to (2.30) with the same constant C which appears from the entropy inequality and is the same constant that is given in (2.16). We will demonstrate how to take the sup of two such functionals. By entropy bound

$$E^{P_N^0}\{F_i\} \le \frac{1}{N}\log E^{P_N}\{\exp(NF_i)\} + C \quad \text{for} \quad i = 1,2\,.$$

If $F = \max(F_1, F_2)$, then

$$(2.33) \qquad E^{P_N^0}[F] \le \frac{1}{N}\log E^{P_N}\{\exp[NF]\} + C$$

and $\exp[NF] \le \exp(NF_1) + \exp(NF_2)$. If

$$\frac{1}{N}\log E^{P_N}\{\exp[NF_i]\} \le 0 \quad \text{for} \quad i = 1,2$$

then

$$\frac{1}{N}\log E^{P_N}\{\exp[NF]\} \le \frac{1}{N}\log 2 \to 0 \quad \text{as} \quad N\to\infty$$

We therefore get

$$E^{P_N^0}\{F\} \le C + o(1) \quad \text{as} \quad N\to\infty$$

Assuming uniform integrability (which we established)

$$E^Q\{F\} \le C\,.$$

We can therefore take the sup over any finite set and by the monotone convergence theorem any arbitrary supremum. This completes the proof of (A).

Proof of (B): To prove (B) we will establish some lemmas.

Lemma 2.3. *For any bounded continuous function $g(x)$ on $\mathcal{R}$ we define*

$$\hat{g}(y) = \frac{1}{M(\lambda)}\int e^{\lambda x - \phi(x)}\,g(x)\,dx$$

where $\lambda = h'(y)$. Then

$$\lim_{k\to\infty}\limsup_{N\to\infty} E^{P_N^0}\left[\frac{1}{N}\int_0^T\sum_{i=1}^N\left|\frac{1}{2k+1}\sum_{|j-i|\le k} g(x_j(t)) - \hat{g}(\bar{x}_{i,k}(t))\right|dt\right] = 0$$

where

$$\bar{x}_{i,k}(t) = \frac{1}{2k+1} \sum_{|j-i|\le k} x_j(t)$$

the mean in a block of size $2k+1$ *centered at* i.

Lemma 2.4.

$$\limsup_{k\to\infty} \limsup_{\epsilon\to 0} \limsup_{N\to\infty} \sup_{0\le r\le N\epsilon} E^{P_N^0} \int_0^T \frac{1}{N} \sum |\bar{x}_{i,k}(t) - \bar{x}_{i+r,k}(t)|\, dt = 0\,.$$

Lemma 2.5. *There exist bounded continuous functions* $g_\ell(x)$ *on* R *which can be used to truncate* $\phi'(x)$ *so that*

$$\limsup_{\ell\to\infty} \limsup_{N\to\infty} E^{P_N^0} \left\{ \frac{1}{N} \int_0^T \sum |g_\ell(x_k(t)) - \phi'(x_i(t))| dt \right\} = 0$$

and

$$\lim_{\ell\to\infty} E^Q \left[\int_0^T \int |\hat{g}_\ell(m(t,\theta)) - h'(m(t,\theta))|\, d\theta \right] = 0\,.$$

Lemma 2.3 is the crux of the entire argument and explains why the answer is what it is. It says that with a very high probability in a typical block around i of size k (large) at a typical time t the values $x_{i-k}, \ldots, x_{i+k}$ appear as if they are drawn as $2k+1$ independent observations from the distribution $\frac{1}{M(\lambda)} e^{\lambda x - \phi(x)}\, dx$ where the parameter λ is picked so that

(2.34)
$$\frac{1}{M(\lambda)} \int x\, e^{\lambda x - \phi(x)}\, dx = \frac{M'(\lambda)}{M(\lambda)} = \bar{x}_{i,k}$$

Inverting (2.33) gives $\lambda = h'(\bar{x}_{i,k})$. Looking at (2.17) we see that we need to study sums like $\frac{1}{N} \sum J''(\frac{i}{N})\, \phi'(x_i(t))$. Since $J(\cdot)$ is a smooth function it is enough to have knowledge of the averages $\frac{1}{2k+1} \sum_{|j-i|\le k} \phi'(x_j(t))$. Lemma 2.5 allows us to pretend that $\phi'(x)$ is bounded and continuous with $\hat{\phi}'(y) = h'(y)$. Lemma 2.4 allows us to replace $\bar{x}_{i,k}$ with k large by $\bar{x}_{i,N\epsilon}$ with ϵ small so that

$$\int_0^T \frac{1}{N} \sum J''(\frac{i}{N})\, \phi'(x_i(t))\, dt$$

can be replaced as $N \to \infty$ by

$$\int \int_0^T J''(\theta)\, h'(m(t,\theta))\, d\theta\, dt$$

Because of (2.21) this will complete the proof of (B).

Before we prove Lemmas 2.3, 2.4 and 2.5 we shall reformulate them in a more convenient form. Our density f_N^t evolves in time and we denote by

$$\bar{f}_N = \frac{1}{T} \int_0^T f_N^t\, dt$$

Then $\bar{f}_N$ is again a probability density on R^N relative to μ_N. Moreover the shift operator τ mapping $\{x_i\}$ to $\{x_{i+1}\}$ on the periodic lattice is a periodic transformation of order N and we can average $\bar{f}_N$ over the orbit $\tau, \tau^2, \ldots, \tau^N$ and denote this average by $\hat{f}_N$. Since

μ_N is τ-invariant this is again a density. If we project this density on a block of size $2k+1$ centered at 0 i.e. $-k \le i \le k$ and obtain a density $\hat{f}_{N,k}$ relative to the product measure μ_{2k+1} on R^{2k+1}. We can also project over two blocks of size $2k+1$ whose centers are separated by τ (at least $2k+1$), and obtain a density $\hat{f}_{N,k,r}$ on $R^{2k+1} \times R^{2k+1}$ relative to $\mu_{2k+1} \times \mu_{2k+1}$. Lemmas 2.3, 2.4 and 2.5 can be restated as Lemmas 2.3′, 2.4′, and 2.5′ in terms of $\hat{f}_{N,k}$ and $\hat{f}_{N,k,r}$.

Lemma 2.3′. *For any bounded continuous function $g(x)$*

$$\lim_{k\to\infty} \limsup_{N\to\infty} \int \left| \frac{g(x_{-k}) + \cdots + g(x_k)}{2k+1} - \hat{g}(\bar{x}_{i,k}) \right| \hat{f}_{N,k}\, d\mu_{2k+1} = 0$$

Lemma 2.4′. *If we denote by x_i, $|i| \le k$ and y_i, $|i| \le k$ the two sets of coordinates on $R^{2k+1} \times R^{2k+1}$,*

$$\lim_{k\to\infty} \limsup_{\epsilon\to 0} \limsup_{N\to\infty} \sup_{2k+1 \le r \le N\epsilon} \int |\bar{x} - \bar{y}|\, \hat{f}_{N,k,r}\, d\mu_{2k+1} \times d\mu_{2k+1} = 0$$

where $\bar{x} = \frac{1}{2k+1}\sum_{|i|\le k} x_i$ and $\bar{y} = \frac{1}{2k+1}\sum_{|i|\le k} y_i$.

Lemma 2.5′. *We can find g_ℓ such that*

$$\lim_{\ell\to\infty} \limsup_{N\to\infty} \int |g_\ell(x) - \phi'(x)|\, \hat{f}_{N,1}(x)\, d\mu_1(x) = 0$$

and

$$\lim_{\ell\to\infty} E^Q \left[\int_0^T \int |\hat{g}_\ell(m(t,\theta)) - h'(m(t,\theta))|\, dt\, d\theta \right] = 0.$$

Except for the second part which requires properties of Q the rest require properties of $\hat{f}_{N,k,r}$ for various N, k, r. These are to be deduced from f_N^0 via f_N^t, $\bar{f}_N$ and $\hat{f}_N$.

Theorem 2.6. *If C is the constant in the entropy bound (2.16) then the same constant C for all N and T*

$$\int \hat{f}_N \log \hat{f}_N\, d\mu_N \le C\, N$$

$$\int W(x)\, \hat{f}_{N,1}(x)\, d\mu_1(x) \le C + \log \int e^{W(x)-\phi(x)}\, dx = C'$$

Proof: By the monotonicity of entropy

$$H(t) = H(f_N^t) = \int f_N^t \log f_N^t\, d\mu_N \le CN \quad \text{for all} \quad t \ge 0\,.$$

By convexity of entropy

$$H(\bar{f}_N) \le \frac{1}{T}\int_0^T H(t)\, dt \le H(0) \le C\, N\,.$$

Again by the convexity of entropy and its invariance under τ due to the invariance of μ_N

$$H(\hat{f}_N) \le H(\bar{f}_N) \le C\, N\,.$$

It is easy to deduce from this that

$$\int \hat{f}_{N,1}(x) \log \hat{f}_{N,1}(x)\, e^{-\phi(x)}\, dx \le C\,.$$

From the entropy inequality the second part follows immediately.

We now need properties of a functional on $L_1(\mu_N)$ related to the Dirichlet form. If f is a probability density on some R^n relative to μ_n we can define for any two indices i,j corresponding to coordinates x_i, x_j in R^n

$$\begin{aligned} I_{i,j}(f) &= \frac{1}{2}\int \frac{1}{f}\cdot\left(\frac{\partial f}{\partial x_i}-\frac{\partial f}{\partial x_j}\right)^2 \mu_n(dx) \\ &= 2\int\left(\frac{\partial\sqrt{f}}{\partial x_i}-\frac{\partial\sqrt{f}}{\partial x_j}\right)^2 \mu_N(dx)\,. \end{aligned}$$

We will mostly be interested in $I_{i,i+1}(f)$ corresponding to nearest neighbor bonds. If f is a density on R^N relative to μ_N and it is projected onto a set A of coordinates containing both i and j and if we denote by f_A the projection then we can compute $I_{i,j}(f)$ as well as $I_{i,j}(f_A)$ and by convexity one can show that

(2.35) $$I_{i,j}(f_A)\le I_{i,j}(f)\quad \text{for all}\quad f$$

Theorem 2.7. *For $\hat{f}_N$ defined before we have*

$$\sum_{i=1}^{N} I_{i,i+1}(\hat{f}_N)\le\frac{C}{N}$$

where C is the same constant in the entropy bound (2.16).

In particular

$$I_{i,i+1}(\hat{f}_N)\le\frac{C}{TN^2}\quad\text{for all}\quad i,\ \text{and } N \text{ and } T\,.$$

Proof: One can show that for the entropy

$$H_N(t)=\int f_N^t\log f_N^t\,\mu_N(dx)$$

$$\begin{aligned} \frac{d}{dt}H_N(t) &= -N^2\textstyle\sum_{i=1}^N I_{i,i+1}(f_N^t) \\ &= -N^2 I^{(N)}(f_N^t) \end{aligned}$$

since H_N and $I^{(N)}$ are nonnegative

$$\int_0^T I^{(N)}(f_N^t)\,dt\le\frac{H_N(0)}{N^2}\le\frac{C}{N}\,.$$

One divides by T and uses convexity. Similarly one averages under the τ orbit and uses convexity. We complete the proofs of Lemmas 2.3′, 2.4′ and 2.5′ by establishing the following theorem.

Let A_{N,C_1,C_2} be the collection of densities f_N on R^N that are τ invariant and satisfy the following bounds

(2.36) $$I_{i,i+1}(f_N)\le\frac{C_1}{N^2}\quad\text{for all}\quad i\,.$$

(2.37) $$\int W(x_1)\,f_N\,d\mu_N=\int W(x)\,f_{N,0}(x)\,e^{-\phi(x)}\,dx\le C_2\,.$$

Theorem 2.8. *For any C_1 and C_2 and bounded continuous $g(\cdot)$*

$$\lim_{k\to\infty} \limsup_{N\to\infty} \sup_{f_N \in A_{N,C_1,C_2}} \int \left| \frac{g(x_k) + \cdots + g(x_k)}{2k+1} - \hat{g}(\bar{x}) \right| f_{N,k}\, d\mu_{2k+1} = 0 \tag{2.38}$$

$$\lim_{k\to\infty} \limsup_{N\to\infty} \limsup_{N\to\infty} \sup_{2k+1 \le r \le N\epsilon} \sup_{f_N \in A_{N,C_1,C_2}} \int |\bar{x} - \bar{y}|\, f_{N,k,r}\, d\mu_{2k+1} \times d\mu_{2k+1} = 0 \tag{2.39}$$

$$\lim_{\ell\to\infty} \sup_{f_N \in A_{N,C_1,C_2}} \int |g_\ell(x) - \phi'(x)|\, f_{N,0}(x)\, d\mu_1(x) = 0 \tag{2.40}$$

for suitable bounded continuous truncations of ϕ'.

$$\lim_{\ell\to\infty} E^Q \left[\int_0^T \int |\hat{g}_\ell(m(t,\theta)) - h'(m(t,\theta))|\, d\theta \right] = 0 \tag{2.41}$$

Proof: Proof of (2.38).

Let us fix k and let $N \to \infty$. By (2.37), $\{f_{N,k}\}$ as f_N varies over A_{N,C_1,C_2} is tight on R^{2k+1}. Let us denote by $B^k_{C_1,C_2}$ the set of limit points. Then we need to show

$$\lim_{k\to\infty} \sup_{f_k \in B^k_{C_1,C_2}} \int \left| \frac{g(x_{-k} + \cdots + g(x_k)}{2k+1} - \hat{g}(\bar{x}) \right| f_k\, d\mu_{2k+1} = 0\ .$$

By (2.37) and (2.38) for all $f_k \in B^k_{C_1,C_2}$

$$I_{i,i+1}(f_k) = 0 \quad \text{for} \quad i, i+1 \subset [-k,k]$$

This implies

$$f_k\, d\mu_{2k+1} = \int \nu_{2k+1,y}(dx)\, \beta(dy)$$

β varies over a class C_k on R

$$\begin{aligned} \int \omega(|x|)\, d\beta &= \int \omega\left(\tfrac{|x_{-k} + \cdots + x_k|}{2k+1}\right) f_k\, d\mu_{k+1} \\ &\le \int \tfrac{1}{2k+1} \left(\omega(|x_{-k}|) + \cdots + \omega(|x_k|)\right) f_k\, d\mu_{k+1} \\ &\le C\ . \end{aligned}$$

Therefore C_k is tight on R. By the law of large numbers that states

$$\lim_{k\to\infty} \int \left| \frac{g(x_{-k} + \cdots + g(x_k)}{2k+1} - \hat{g}(y) \right| d\nu_{k,y} \to 0$$

uniformly for y in compact sets (2.38) follows.

Proof of (2.39): Let us take again a limit point f now in $R^{2k+1} \times R^{2k+1}$ of $f_{N,k,r}$ as $N \to \infty$ and r behaves in any manner. The limit is a measure on $R^{2k+1} \times R^{2k+1}$ with x and y coordinates. The $I_b(f)$ of bonds $i, i+1$ among the x coordinates are 0 and $I_b(f)$ of bonds among the y coordinates are zero. The only problem is the bond between x and y. Let us take such a bond δ. Then

$$I_\delta(f) \le \liminf_{N\to\infty} I_\delta(f_N)\ .$$

But the δ bond does not exist initially. But by telescoping neighboring bonds we can estimate it by

$$I_\delta(f_N) \le (N\epsilon)^2 I_{i,i+1}(f_N) \le (N\epsilon)^2 \frac{C}{TN^2} \le \frac{\epsilon^2}{T} .$$

If we take the ϵ limit as well, $I_\delta(f) = 0$. So our $\bar{x}$ and $\bar{y}$ behave like averages of two half blocks in a block of size $2k+2$ under $\nu_{2k+2,y}$ and are nearly equal. The bound

$$\int W(x)\, f_{N,0}(x)\, e^{-\phi(x)}\, dx \le C$$

gives enough uniform integrability that the unboundedness of $|x|$ is no problem. This proves (2.39).

Proof of (2.40): Again because the inequality

$$\int \omega\left(|\phi'|(x)\right) f_{N,0}(x)\, d\mu_1(x) \le \int W(x)\, f_{N,0}(x)\, d\mu_1(x) \le C$$

gives enough control that the standard truncation

$$\begin{aligned} g_\ell(x) &= \phi'(x) \quad \text{if} \quad \phi'(x) \in [-\ell, \ell] \\ &= \ell \quad \text{if} \quad l\phi'(x) > \ell \\ &= -\ell \quad \text{if} \quad \phi'(x) < -\ell \end{aligned}$$

works.

Proof of (2.41): Using entropy bounds one can get an inequality of the form

$$|\hat{g}_\ell(y)| \le A + Bh(y)$$

and

$$|h'(y)| \le A + B\, h(y)$$

(2.41) then follows from (2.32).

Finally we now turn to the proof of (C).

Consider the expectation

$$\begin{aligned} &\textstyle\int \sum J(\frac{i}{N}) [\phi'(x_i) - \phi'(x_{i+1})]\, f_N\, \mu_N(dx) \\ &= \textstyle\int \sum J(\frac{i}{N}) [\phi'(x_i) - \phi'(x_{i+1})]\, f_N\, e^{-\sum \phi(x_i)}\, dx \\ &= \textstyle\int \sum J(\frac{i}{N}) \left(\frac{\partial f_N}{\partial x_i} - \frac{\partial f_N}{\partial x_{i+1}}\right) e^{-\sum \phi(x_i)}\, dx \\ &\le \left(\textstyle\int \frac{1}{N} \sum J^2(\frac{i}{N})\, f_N\, e^{-\sum \phi(x_i)}\, dx\right)^{1/2} \sqrt{2 I_N(f)} \\ &\le \sqrt{2C} \left\{E^Q \left[\textstyle\int J^2(\theta)\, d\theta\right]\right\}^{1/2} \quad \text{as} \quad N \to \infty \end{aligned}$$

On the other hand

$$\begin{aligned} &\textstyle\int \sum J(\frac{i}{N}) [\phi'(x_i) - \phi'(x_{i+1})]\, f_N\, \mu_N(dx) \\ &= \textstyle\int \sum \left\{J(\frac{i}{N}) - J(\frac{i-1}{N})\right\} \phi'(x_i)\, f_N\, \mu_N(dx) \\ &= \textstyle\frac{1}{N} \int \sum J'(\frac{i}{N})\, \phi'(x_i)\, f_N\, \mu_N(dx) \\ &= E^Q \textstyle\int J'(\theta)\, h'(m(\theta))\, d\theta \end{aligned}$$

Here Q is the weak limit of

$$\frac{1}{N} \sum x_i \delta_{i/N}$$

as $N \to \infty$. We assume a bound of the form

$$\int \frac{1}{N} \sum W(x_i)\, f_N \, d\mu_N \leq C$$

to justify all we did. We then get an estimate

$$E^Q \left[\int J'(\theta)\, h'(m(\theta))\, d\theta \right] \leq \left(E^Q \int J^2(\theta)\, d\theta \right)^{1/2} \cdot C$$

We can now pick $J(\theta)$ to depend on variables $z_1, \ldots, z_k$ where

$$z_i = \frac{1}{N} \sum G_i(\frac{k}{N}) x_k \, .$$

The integration by parts generates additional terms that go to zero as $N \to \infty$. We get in the end

$$E^Q \left[\int J'(\theta, z_1, \ldots, z_k)\, h'(m(\theta))\, d\theta \right] \leq C \left(E^Q \int J^2 \theta, z_1, \ldots, z_k)\, d\theta \right)^{1/2}$$

where $z_i = \int G_i(\theta)\, m(\theta)\, d\theta$. Since such functions $J(\theta, z_1, \ldots, z_k)$ are dense we have essentially

$$E^Q \left[\int J'(\theta, \omega)\, h'(m(\theta))\, d\theta \right] \leq C \cdot \left(E^Q \int J^2(\theta, \omega)\, d\theta \right)^{1/2}$$

where $\omega = m(\cdot)$. This is essentially (C) stated in dual form.

3 Large Deviations and Weak Asymmetries.

The hydrodynamic limit that we proved in the last lecture is really a form of weak law of large numbers. If we start with an initial density f_N^0 having a microscopic initial profile $m_0(\theta)\, d\theta$, satisfying in addition a linear entropy bound as $N \to \infty$, then the measure Q_N^0 (induced by the process P_N^0) on the space $\Omega = C([0,T], \mathcal{M}]$ of signed measure valued variables on $[0,T]$ converges weakly as $N \to \infty$, to the degenerate distribution on the trajectory $m(t,\theta)\, d\theta$ in Ω, $0 \leq t \leq T$ where $m(t,\theta)$ solves the nonlinear heat equation.

$$\begin{aligned} \frac{\partial m}{\partial t} &= \tfrac{1}{2}(h'(m))_{\theta\theta}\, , \quad t > 0 \\ m(t,\cdot)|_{t=0} &= m_0(\cdot) \end{aligned} \tag{3.1}$$

One can ask the rate of decay of the probabilities under Q_N^0 for sets away from the above solution trajectory. This problem is interesting even if we start from equilibrium, i.e. we choose $f_N^0 \equiv 1$ so that $Q_N^0 \equiv Q_N$ and $P_N^0 = P_N$ the stationary process. We seek to prove a large deviation theorem of the form

$$\limsup_{N \to \infty} \frac{1}{N} \log Q_N(C) \leq - \inf_{\omega \in C} B(\omega) \tag{3.2}$$

for closed sets and

$$\liminf_{N \to \infty} \frac{1}{N} \log Q_N(G) \geq - \inf_{\omega \in G} B(\omega) \tag{3.3}$$

for open sets G.

Before we outline the proof let us see if we can guess the correct rate function $B(\omega)$. Suppose we take a function $J(t,\theta)$ on $[0,T]\times S$ that is smooth and write

$$(3.4)\qquad \begin{aligned} d\sum J(t,\tfrac{i}{N})x_i(t) &= \sum J_t(t,\tfrac{i}{N})\,x_i(t)\,dt \\ &\quad +\tfrac{1}{2}\sum J_{\theta\theta}(t,\tfrac{i}{N})\,\phi'(x_i(t))\,dt \\ &\quad +\sum J_\theta(t,\tfrac{i}{N})\,d\beta_{i,i+1}(t) \\ &\quad + \text{negligible terms,} \end{aligned}$$

we can then write for each J an exponential martingale of the form

$$\exp\{\psi_N(t,\omega) + \text{negligible terms}\}$$

where

$$\begin{aligned} \psi_N(t,\omega) &= \int_0^t \sum J_\sigma(\sigma,\tfrac{i}{N})\,x_i(\sigma)\,d\sigma \\ &\quad +\tfrac{1}{2}\int_0^t \sum J_{\theta\theta}(\sigma,\tfrac{i}{N})\,\phi'(x_i(\sigma))\,d\sigma \\ &\quad -\tfrac{1}{2}\int_0^t \sum \left[J_\theta(\sigma,\tfrac{i}{N})\right]^2\,d\sigma \\ &\quad +\sum J(t,\tfrac{i}{N})\,x_i(t) - \sum J(0,\tfrac{i}{N})\,x_i(0) \end{aligned}$$

If we multiply $\psi_N(T,\omega)$ by $\exp(\sum G(\frac{i}{N})\,x_i(0))$ and integrate with respect to P_N we get

$$(3.5)\qquad \begin{aligned} &\tfrac{1}{N}\log E^{P_N}\left\{\exp\left\{\psi_N(T,\omega)+\sum G(\tfrac{i}{N})\,x_i(0)\right\}\right\} \\ &\quad \simeq \tfrac{1}{N}\sum \log M(\lambda(G(\tfrac{i}{N}))) \end{aligned}$$

The expression $\psi_N(T,\omega)$ can be reexpressed in terms of $\omega=\nu(\cdot,d\theta)$ as

$$\begin{aligned} \tfrac{1}{N}\psi_N(T,\omega) &= \int J(T,\theta)\,\nu(T,d\theta) - \in J(0,\theta)\,\nu(0,d\theta) \\ &\quad -\int_0^T\int J_t(t,\theta)\,\nu(t,d\theta)\,dt \\ &\quad -\text{"difficult term"} \\ &\quad -\tfrac{1}{2}\int_0^T\int J^2(t,\theta)\,d\theta\,dt \end{aligned}$$

If we keep in mind that as $N\to\infty$ we replaced the difficult term by

$$\frac{1}{2}\int_0^T\int h'(m(t,\theta))\,J_{\theta\theta}(t,\theta)\,d\theta\,dt$$

in our earlier lecture from Tchebichev's inequality we get a large deviation upper bound that is equal to

$$B(\omega)=\sup_{J,G}\left[\hat{\psi}_{G,J}(\omega)\right]$$

where after integrating by parts

$$\begin{aligned} \hat{\psi}_{J,G}(\omega) &= \int\int J(t,\theta)\,m_t(t,\theta)\,d\theta\,dt \\ &\quad -\tfrac{1}{2}\int\int J(t,\theta)\,(h'(m(t,\theta)))_{\theta\theta}\,d\theta\,dt \\ &\quad +\int G(\theta)\,m(0,\theta)\,d\theta \\ &\quad -\tfrac{1}{2}\int\int J^2(t,\theta)\,d\theta\,dt\,. \end{aligned}$$

We see that

$$B(\omega)=\int h(m(0,\theta))\,d\theta+\frac{1}{2}\int_0^T\|\frac{\partial m}{\partial t}-\frac{1}{2}(h'(m))_{\theta\theta}\|_{-1}^2\,dt$$

To justify this we have to replace the "difficult term"

$$\int_0^T \frac{1}{N}\sum J_{\theta\theta}(t,\frac{i}{N})\,\phi'(x_i(t))\,dt \quad \text{by} \quad \int_0^T \int J_{\theta\theta}(t,\theta)\,h'(m(t,\theta))\,d\theta\,dt$$

with error probabilities that are superexponentially small as $N \to \infty$. After ignoring truncation difficulties which are again handled with superexponential controls using entropy bounds this amounts to showing that for any bounded $g(x)$ on R which is continuous

$$E^{P_N}\left\{\exp \int_0^T \sum \left| \frac{g(x_{i-N\epsilon}(t)) + \cdots + g(x_{i+N\epsilon}(t))}{2N\epsilon + 1} - \hat{g}\left(\frac{x_{i-N\epsilon}(t) + \cdots + x_{i+N\epsilon}(t)}{2N\epsilon}\right)\right| dt\right\} \le \exp[C_\epsilon N]$$

where $C_\epsilon \to 0$ as $N \to \infty$. By using Feynman-Kac formula this is related to an eigenvalue which can be estimated by a variational formula. We can again truncate high values of x_i effectively and reduce it to proving

$$\lim_{\epsilon\to 0}\limsup_{N\to\infty} \sup_{\substack{f_N:\, \frac{1}{N}\int \sum W(x_i) f_N\, d\mu_N \le C_1 \\ I_N(f_N) \le C_2/N}} \int \frac{1}{N}\sum |g^\epsilon_{i,N} - \hat{g}^\epsilon_{i,N}|\, f_N\, d\mu_N = 0$$

where

$$g^\epsilon_{i,N} = \frac{1}{2N\epsilon + 1}\sum_{|j-i|\le N\epsilon} g(x_j)$$

and

$$\hat{g}^\epsilon_{i,N} = \hat{g}\left(\frac{1}{2N\epsilon + 1}\sum_{|j-i|\le N\epsilon} x_j\right)$$

This was accomplished in the last lecture.

This proves the upper bound except that we need superexponential tightness estimates on Q_N. These take the form

$$\limsup_{\ell\to\infty}\limsup_{N\to\infty}\frac{1}{N}\log P_N\left\{\sup_{0\le t\le T}\frac{1}{N}\sum|x_i(t)| \ge \ell\right\} = -\infty \tag{3.6}$$

$$\limsup_{\delta\to 0}\limsup_{N\to\infty}\frac{1}{N}\log P_N\left\{\sup_{\substack{0\le s<t\le T\\ (t-s)\le\delta}}\frac{1}{N}\int_s^t \sum|\phi'(x_i(t))|\,dt \ge \epsilon\right\} = -\infty \tag{3.7}$$

$$\limsup_{\delta\to 0}\limsup_{N\to\infty}\frac{1}{N}\log P_N\left\{\sup_{\substack{0\le s<t\le T\\ (t-s)\le\delta}}\left|\frac{1}{N}\sum_i \int_s^t J'(\frac{i}{N})\,d\beta_{i,i+1}(\sigma)\right| \ge \epsilon\right\} = -\infty \tag{3.8}$$

We derived (3.6) and (3.7) in the last lectures. Although we did not need superexponential estimates we needed them for P_N^0 which we controlled via entropy bounds after deriving them for P_N in the superexponential form. As for (3.8) the superexponential estimates are readily obtained because it is essentially for Brownian motion.

We now turn to the lower bound. We need to study what is known as weak asymmetries. We change the evolution operator to

$$\hat{L}_N = L_N + \sum b_{i,i+1}(t, \frac{i}{N}, x)\left(\frac{\partial}{\partial x_i} - \frac{\partial}{\partial x_{i+1}}\right)$$

Here $b(t,\theta,x)$ is a local function of x that is bounded continuous and depends smoothly on t and θ. The local dependence is on a finite number of coordinates x_i for $|i| \le k_0$. If we denote by τ_i the shift in coordinates from $x_\alpha \to x_{i+\alpha}$ then

$$b_{i,i+1}(t, \frac{i}{N}, x) = b(t, \frac{i}{N}, \tau_i x)$$

We denote by $\hat{P}_N^0$ the diffusion process for $\hat{L}_N$ starting from some f_N^0 with linear entropy bounds in N. Let Q_N^0 denote the induced measure on the space Ω. We will need the function

$$\hat{b}(t,\theta,y) = E_y\{b(t,\theta,x)\}$$

where E_y refers to the measure $\prod \frac{1}{M(\lambda)} e^{\lambda x_i - \phi(x_i)} dx_i$ with $\lambda = h'(y)$. We will only need a few coordinates depending on specific b that we use.

Using Girsanov's formula for the Radon-Nikodym derivative of $\hat{P}_N^0$ to P_N^0 we can calculate the entropy of $\hat{P}_N$ relative to P_N and in addition to the entropy of f_N^0, we pick up an additional term from Girsanov's formula that is again bounded by CTN the constant C coming from the bound on b. Because under P_N the "difficult term" gets replaced by $\int\int J_{\theta\theta}(t,\theta)\, h'(m(t,\theta))\, d\theta\, dt$ with superexponential error we can also replace it under $\hat{P}_N^0$. A similar argument allows us to replace expressions of the form

$$\frac{1}{N}\int_0^T \sum J(t, \frac{i}{N})\, b_{i,i+1}(t, \frac{i}{N}, x(t))\, dt$$

by

$$\int_0^T \int J(t,\theta)\, \hat{b}(t,\theta,m(t,\theta))\, d\theta\, dt$$

under $\hat{P}_N^0$. If we now calculate

$$\begin{aligned} d(\tfrac{1}{N}\sum J(\tfrac{i}{N})\, x_i(t)) &= \tfrac{1}{N}\sum J_{\theta\theta}(\tfrac{i}{N})\, \phi(x_i(t))\, dt \\ &\quad -\tfrac{1}{N}\sum J_\theta(\tfrac{i}{N})\, b_{i,i+1}(t, \tfrac{i}{N}, x(t))\, dt \\ &\quad + \text{negligible terms,} \end{aligned}$$

we end up with the equation

$$\frac{\partial m}{\partial t} = \frac{1}{2}(h'(m))_{\theta\theta} + (\hat{b}(t,\theta,m(t,\theta)))_\theta$$

If we have a uniqueness theorem for solutions of the above equation (in weak form) then we obtain a hydrodynamic limit for $\hat{P}_N^0$.

Of special interest is the choice of $b(t,\theta,x) = c(t,\theta)$ independent of x. In such a case the equation is

$$\frac{\partial m}{\partial t} = \frac{1}{2}(h'(m))_{\theta\theta} + \frac{\partial}{\partial\theta}c(t,\theta)\,.$$

The entropy contribution of the Ginsanov term equals

$$\frac{1}{2}\int_0^T \int c^2(t,\theta)\,dt\,d\theta\ .$$

This allows us to give a lower bound for the large deviation. If we pick f_N^0 to be of the form

$$f_N^0 = \prod \frac{1}{M(\lambda(\frac{i}{N}))} e^{\sum \lambda(i/N) x_i}$$

where $\lambda(\theta) = h'(m_0(\theta))$, then the entropy of $\hat{P}_N^0$ relative to P_N is roughly

$$N\left[\int h(m_0(\theta))\,d\theta + \frac{1}{2}\int_0^T \int [c(t,\theta)]^2\,dt\,d\theta\right] = B(\omega)$$

for $\omega = m(\cdot,\cdot)\,d\theta$. To make it come out we have to pick $c(t,\theta)$ so that $\int c(t,\theta)\,d\theta \equiv 0$ for all t. We can always do this for the equation only involving $\partial c/\partial\theta$.

4 The Method of Relative Entropy.

We will explain the method of relative entropy in the special case that we have studied using Dirichlet form techniques. We start with a solution of our nonlinear heat equation

$$\frac{\partial m}{\partial t} = \frac{1}{2}(h'(m))_{\theta\theta}$$

which is very regular. Then we define the conjugate variable $\lambda(t,\theta) = h'(m(t,\theta))$. We take as our initial density a very special f_N^0 given by

$$f_N^0 = \exp\left[\psi_N^0(x_1,\ldots,x_N)\right]$$

where

$$\psi_N^0(x_1,\ldots,x_N) = \sum \lambda(0,\frac{i}{N})x_i - \sum \log m(\lambda(0,\frac{i}{N}))$$

We are hoping that the density at time t will be given by

$$g_N^t = \exp\left[\psi_N^t(x_1,\ldots,x_N)\right]$$

of the same form where

$$\psi_N(t,x_1,\ldots,x_N) = \sum \lambda(t,\frac{i}{N})x_i - \sum \log M(\lambda(t,\frac{i}{N})) \tag{4.1}$$

In reality of course f_N^t has to be obtained as the solution of

$$\frac{\partial f_N^t}{\partial t} = L_N f_N^t$$

$$f_N^t|_{t=0} = f_N^0 = g_N^0 = \exp[\psi_N^0]\ .$$

f_N^t and g_N^t will diverge from each other as t becomes positive, but we will show that

Theorem 4.1. *For any finite interval* $[0,T]$

$$\lim_{N\to\infty}\frac{1}{N}\sup_{0\le t\le T}\int f_N^t\log\frac{f_N^t}{g_N^t}\,d\mu_N=0\,.$$

Remark. By construction g_N^t has the right profile at t namely $m(t,\cdot)$. Moreover by large deviation theory for sums of independent random variables the probability of deviations from the correct profile under g_N^t are exponentially small. However because f_N^t has sublinear entropy relative to g_N^t the probability of deviation from the profile goes to zero under f_N^t thereby establishing the hydrodynamic scaling limit.

The method of proof involves looking at

$$H_N(t)=\frac{1}{N}\int f_N^t\log\frac{f_N^t}{g_N^t}\,d\mu_N \tag{4.2}$$

and calculating $\frac{dH_N(t)}{dt}$. We show that for some constant C

$$\frac{dH_N(t)}{dt}\le CH_N(t)+\epsilon_N(t)$$

and

$$\lim_{N\to\infty}\int_0^T\epsilon_N(t)\,dt=0\,.$$

From this

$$\begin{aligned}\frac{dH_N(t)e^{-Ct}}{dt}&=e^{-ct}\frac{dH_N(t)}{dt}-CH_N(t)\,e^{-Ct}\\&\le e^{-Ct}\,\epsilon_N(t)\,.\end{aligned}$$

Since $H_N(0)=0$, by choice it follows that

$$H_N(t)\to 0 \text{ uniformly on } [0,T]\,.$$

Lemma 4.2. *With $H_N(t)$ defined by (4.2)*

$$\frac{dH_N(t)}{dt}\le\frac{1}{N}\int\frac{L_Ng_N^t-\frac{\partial}{\partial t}g_N^t}{g_N^t}\cdot f_N^t\,d\mu_N\,.$$

Proof: Denoting $\frac{\partial f}{\partial t}$ by $\dot f$, we calculate

$$\frac{dH_N(t)}{dt}=\frac{1}{N}\int\dot f_N^t\log\frac{f_N^t}{g_N^t}\,d\mu_N+\frac{1}{N}\int\dot f_N^t\,d\mu_N-\frac{1}{N}\int\frac{\dot g_N^t}{g_N^t}\,f_N^t\,d\mu_N$$

We use the relations

$$\dot f_N^t=L_N^*f_N\,,\qquad\int\dot f_N^t\,d\mu_N=\frac{d}{dt}1=0\,,$$

$$\begin{aligned}\frac{dH_N(t)}{dt}&=\frac{1}{N}\int L_N^*f_N^t\log\frac{f_N^t}{g_N^t}\,d\mu_N-\frac{1}{N}\int\frac{\dot g_N^t}{g_N^t}\,f_N^t\,d\mu_N\\&=\frac{1}{N}\int f_N^t\,L_N\log\frac{f_N^t}{g_N^t}\,d\mu_N-\frac{1}{N}\int\frac{\dot g_N^t}{g_N^t}\,f_N^t\,d\mu_N\end{aligned}$$

We now use the maximum principle for L_N in the form $L_N\psi(u) \le \psi'(u)\, L_N u$ for concave ψ. We use it for $\psi(u) = \log u$

$$
\begin{aligned}
\frac{dH_N(t)}{dt} &\le \tfrac{1}{N}\int f_N^t \cdot \tfrac{g_N^t}{f_N^t} \cdot L_N \tfrac{f_N^t}{g_N^t}\, d\mu_N - \tfrac{1}{N}\int \tfrac{\dot g_N^t}{g_N^t} f_N^t\, d\mu_N \\
&= \frac{1}{N}\int g_N^t\, L_N \frac{f_N^t}{g_N^t}\cdot d\mu_N - \frac{1}{N}\int \frac{\dot g_N^t}{g_N^t} f_N^t\, d\mu_N \\
&= \frac{1}{N}\int \frac{L_N^* g_N^t}{g_N^t}\cdot f_N^t\, d\mu_N - \frac{1}{N}\int \frac{\dot g_N^t}{g_N^t} f_N^t\, d\mu_N \\
&= \frac{1}{N}\int \frac{L_N^* g_N^t - \dot g_N^t}{g_N^t} f_N^t\, d\mu_N
\end{aligned}
$$

which proves the lemma. We have used L_N^* although in our case $L_N = L_N^*$. But the lemma is valid without the symmetry assumption.

The next step is to calculate explicitly

$$
\frac{1}{N}\cdot\frac{1}{g_N^t}\left(L_N g_N^t - \frac{\partial}{\partial t} g_N^t\right) = B_N(t, x_1, \ldots, x_N)
$$

using the representation $g_N^t = \exp[\psi_N^t(x_1, \ldots, x_N)]$ where ψ_N is given by (4.1).

$$
\begin{aligned}
\frac{1}{N}\frac{1}{g_N^t} L_N g_N^t &= \tfrac{1}{N} L_N \psi_N + \tfrac{N}{2}\sum\left(\tfrac{\partial\psi_N}{\partial x_i} - \tfrac{\partial\psi_N}{\partial x_{i+1}}\right)^2 \\
&= \frac{1}{N}\cdot\frac{N^2}{2}\sum\Big(\lambda(t, \frac{i+1}{N}) - 2\lambda(t, \frac{i}{N}) + \lambda(t, \frac{i-1}{N})\Big)\,\phi'(x_i) \\
&\quad + \frac{N}{2}\sum\left[\lambda(t, \frac{i+1}{N}) - \lambda(t, \frac{i}{N})\right]^2
\end{aligned} \tag{4.3}
$$

$$
\begin{aligned}
\frac{1}{N}\frac{1}{g_N^t}\frac{\partial g_N^t}{\partial t} &= \tfrac{1}{N}\cdot\tfrac{\partial}{\partial t}\psi_N = -\tfrac{1}{N}\sum \tfrac{M'(\lambda(t,i/N))}{M(\lambda(t,i/N))}\lambda_t(t, \tfrac{i}{N}) + \tfrac{1}{N}\sum \lambda_t(t, \tfrac{i}{N}) x_i \\
&= \frac{1}{N}\sum \lambda_t(y, \frac{i}{N})(x_i - m(t, \frac{i}{N}))
\end{aligned} \tag{4.4}
$$

We have used the relation $\lambda = h'(m)$ which implies $m = \dfrac{M'(\lambda)}{M(\lambda)}$. Moreover

$$
\frac{d\lambda}{dt} = \frac{d\lambda}{dm}\cdot\frac{dm}{dt} = h''(m)\frac{1}{2}\lambda_{\theta\theta}
$$

Moreover we can do summation by parts and write

$$
\begin{aligned}
&-\frac{1}{2N}\sum\left[\lambda(t, \frac{i+1}{N}) - \lambda(t, \frac{i}{N})\right]^2 \\
&= \frac{1}{2N}\sum\left[\lambda(t, \frac{i+1}{N}) - 2\lambda(t, \frac{i}{N}) + \lambda(t, \frac{i-1}{N})\right]\, \lambda(t, \frac{i}{N}) \\
&= \frac{1}{2N}\sum(\nabla^2\lambda)(t, \frac{i}{N})\, h'(m(t, \frac{i}{N}))
\end{aligned} \tag{4.5}
$$

Combining (4.2), (4.3), (4.4) and (4.5)

$$
\begin{aligned}
B_N(t, x_1, \ldots, x_N) &= \frac{N^2}{2N}\sum\left(\nabla^2\lambda(t, \frac{i}{N})\right)\left(\phi'(x_i) - h'(m(t, \frac{i}{N}))\right) \\
&\quad - \frac{1}{2N}\sum \lambda_{\theta\theta}(t, \frac{i}{N})\, h''(m(t, \frac{i}{N}))(x_i - m(t, \frac{i}{N}))
\end{aligned}
$$

Step 1. $N^2(\nabla^2\lambda)(t,\frac{i}{N}) = \lambda_{\theta\theta}(t,\frac{i}{N}) + \epsilon_N$.

To do this we need control on $\phi'(x_i)$ and f_N^t has entropy bounds relative to μ_N. Therefore

$$E^{f_N^t}\frac{1}{N}\sum|\phi'(x_i)| \text{ is controlled.}$$

Step 2. Replace $\phi'(x_i)$ by $\xi_{k,i} = \dfrac{1}{2k+1}\sum_{|j-i|\le k}\phi'(x_j)$ and x_i by $\bar{x}_{i,k} = \frac{1}{2k+1}\sum_{|j-i|\le k} x_j$. This introduces errors which are small but their smallness depends on k. We denote these errors by $\epsilon_{N,k}$. We need control on

$$\int \frac{1}{N}\sum |x_i|\, f_N^t\, d\mu_N$$

and

$$\int \frac{1}{N}\sum |\phi'(x_i)|\, f_N^t\, d\mu_N$$

which are again obtained from entropy.

We then replace $\xi_{i,k}$ by $h'(\bar{x}_{i,k})$. This requires some results obtained earlier using Dirichlet forms. But this is only for long microscopic blocks. We still have to handle some truncation. But we will ignore this complication. Finally we have replaced $B_N(t,x_1,\dots,x_N)$ by $C_N(t,x_1,\dots,x_N)$ given by

$$\begin{aligned} C_N^k(t,x_1,\dots,x_N) &= \frac{1}{2N}\sum\lambda_{\theta\theta}(t,\frac{i}{N})\Big[h'(\bar{x}_{i,k}) - h'(m(t,\frac{i}{N})) \\ &\quad -(\bar{x}_{i,k} - m(t,\frac{i}{N}))\,h''(m(t,\frac{i}{N}))\Big] \end{aligned} \tag{4.6}$$

such that

$$\int_0^T\int B_N(t,x_1,\dots,x_N)\, f_N^t\, d\mu_N\, dt \le \int_0^T\int C_N^k(t,x_1,\dots,x_N)\, f_N^t\, d\mu_N + \epsilon_{N,k} + \delta_k \tag{4.7}$$

where $\epsilon_{N,K}\to 0$ as $N\to\infty$ for each fixed k and $\delta_k\to 0$ as $k\to\infty$. If we use Taylor's expansion

$$|h'(\bar{x}) - h'(m) - (\bar{x}-m)\,h''(m)| \le C|\bar{x}-m|^2\,.$$

In fact by truncation which we can perform and which we have decided to ignore

$$|h'(\bar{x}) - h'(m) - (\bar{x}-m)\,h''(m)| \le K(|\bar{x}-m|) \tag{4.8}$$

where $K(|x|)$ is a bounded function behaving near 0 with an estimate

$$K(|x|) \le C|x|^2\,. \tag{4.9}$$

From (4.6), (4.7) and (4.8)

$$\begin{aligned} &\int_0^T\int B_N(t,x_1,\dots,x_N)\, f_N^t\, d\mu_N\, dt \\ &\le C_0\frac{1}{N}\int\sum K(|\xi_{i,k} - m(t,\frac{i}{N})|)\, f_N^t\, d\mu_N + \epsilon_{N,k} + \delta_k \end{aligned} \tag{4.9}$$

If we use entropy inequality we get

$$
\begin{aligned}
&\frac{1}{N}\int \sum K(|\xi_{i,k} - m(t,\frac{i}{N})|)\, f_N^t\, d\mu_N \\
(4.10) \qquad &\leq \frac{1}{N\rho}\log\int \exp\left\{\rho \sum K(|\xi_{i,k} - m(t,\frac{i}{N})|)\right\} g_N^t\, d\mu_N + \frac{1}{\rho}H_N(t)
\end{aligned}
$$

By large deviation theory

$$
\begin{aligned}
&\lim_{k\to\infty}\lim_{N\to\infty}\frac{1}{N}\log\int \exp\left\{\rho \sum K(|\xi_{i,k} - m(t,\frac{i}{N})|)\right\} g_N^t\, d\mu_N \\
(4.11) \qquad &= sup_{\tilde{m}(\cdot)}\left[\rho\int K(|\tilde{m}(\theta) - m(t,\theta)|)\, d\theta - \int R(m(t,\theta),\tilde{m}(\theta))\, d\theta\right]
\end{aligned}
$$

for a rate function $R(t,\tilde{m})$ having the property $R(m,\tilde{m}) \to \infty$ as $|\tilde{m}| \to \infty$ and $R(m,\tilde{m}) \geq \delta(m-\tilde{m})^2$ for m in a compact set. If ρ is small enough

$$
\rho K(|m - \tilde{m}|) \leq R(m,\tilde{m})
$$

for all $\tilde{m}$ and m of the form $m = m(t,\theta)$ for some $t \in [0,T]$ and θ. Therefore for such a ρ the supremum in (4.11) is zero and using (4.10) and (4.11)

$$
\lim_{k\to\infty}\limsup_{N\to\infty}\int_0^T\int B_N(t,x_1,\ldots,x_N)\, f_N^t\, d\mu_N\, dt \leq 0
$$

But our expression B_N does not depend on k and we are done.

5 Hamiltonian Systems and Euler's Equations.

One of the problems of kinetic theory is the transition from a Hamiltonian description of a large system of particles to a fluid description governed by Euler's equation. This is usually done through the intermediary of Boltzmann equations and is not totally satisfactory.

We will in this lecture attempt a direct approach from particles to fluids. But let me warn you that we will fail. We will have to make several modifications before we prove a mathematical result. To save time let us start already with a scaled Hamiltonian system. The physical space is the 3-torus T^3. The phase space is $T^3 \times R^3$. The positions of particles will be denoted by q^α and their velocities by p^α. The components will be q_i^α and p_i^α, $i = 1,2,3$. α will identify the particle. There will be N of them $\alpha = 1,2,\ldots,N$. The scale parameter ϵ will be related to N by $N = \epsilon^{-3}$. The number ϵ is the effective range of interaction between particles. N will go to ∞ and $\epsilon \to 0$, always related by $N = \epsilon^{-3}$. We will therefore freely use ϵ or N as is convenient. The interaction will be through a pair potential $V(\frac{q^\alpha - q^\beta}{\epsilon})$ between two particles located at q^α and q^β. We shall assume that V has compact support. Let us assume for simplicity that $V \geq 0$ and $V(0) > 0$. The Hamiltonian energy is

$$
H = \frac{1}{2}\sum_\alpha\sum_i |p_i^\alpha|^2 + \frac{1}{2}\sum_{\alpha,\beta} V(\frac{q^\alpha - q^\beta}{\epsilon})
$$

The equations of motion are

$$\frac{dq_i^\alpha}{dt} = \frac{\partial H}{\partial p_i^\alpha}$$
$$\frac{dp_i^\alpha}{dt} = -\frac{\partial H}{\partial q_i^\alpha}$$

There are five conserved quantities for our system. The number of particles is conserved. In addition we have the three components of momenta and the energy itself. Corresponding to these we have their spatial distributions viewed as signed measures on T^3.

$$\begin{aligned}\xi_0(dx,t) &= \frac{1}{N}\sum \delta(q^\alpha(t))\\ \xi_i(dx,t) &= \frac{1}{N}\sum \delta(q^\alpha(t))\, p_i^\alpha(t)\,, \quad i=1,2,3\\ \xi_4(dx,t) &= \frac{1}{N}\sum \delta(q^\alpha(t))\, h^\alpha\end{aligned}$$

where

$$h^\alpha = \frac{1}{2}\sum_{i=1}^{3}|p_i^\alpha(t)|^2 + \frac{1}{2}\sum_\beta V\left(\frac{q^\alpha(t)-q^\beta(t)}{\epsilon}\right)$$

Under certain conditions to be discovered, as $N\to\infty$, these quantities are supposed to have limits

$$\xi_i(dx,t)\to u_i(x,t)\,dx$$

in the weak sense and the five quantities u_i, $i=0,1,2,3,4$ are expected to satisfy the Euler equation

$$\frac{\partial u}{\partial t} + \nabla_x F(u) = 0\,.$$

$F = F_{ij}$ is a 5×3 matrix of functions of $u_0,\ldots,u_4$ and the equations are a system of conservation laws

$$\frac{\partial u_i}{\partial t} + \sum_{j=1}^{3}\frac{\partial}{\partial x_j}F_{ij}(u) = 0\,, \qquad 0\le i<4$$

The derivation of these equations, even at a formal level, requires familiarity with the thermodynamical formalism and we will not attempt it here.

If we take a finite volume Q in R^3 and consider the union of all $(Q\times R^3)$

$$S = \bigcup_{n=0}^{\infty}(Q\times R^3)^n$$

we can define a probability measure on S indexed by parameters λ_i, $0\le i\le 4$; $\lambda_0,\lambda_1,\lambda_2$, λ_3 will be real and λ_4 will be positive.

On $(Q\times R^3)^n$ the measure will have a density relative to Lebesgue measure given by

$$\frac{1}{Z}\frac{1}{n!}\exp\left[\lambda_0 n + \sum_\alpha\sum_{i=1}^{3}\lambda_i p_i^\alpha - \lambda_4\left\{\frac{1}{2}\sum_\alpha\sum_i (p_i^\alpha)^2 + \frac{1}{2}\sum_\alpha\sum_\beta V(q^\alpha - q^\beta)\right\}\right]$$

Z is a normalization constant depending on Q. The quantity

$$\lim_{Q\uparrow R^3}\frac{1}{|Q|}\log Z = \psi(\lambda)$$

plays an important role in the theory. We need only to know that there is a good set G of possible values for λ in R^5 such that for λ in the set as $Q \uparrow R^3$ our measures on $\cup_n (Q \times R^3)^n$ have limits as stationary point process on $R^3 \times R^3$. The locations q^α is a stationary point process on R^3 depending on the parameters λ_0 and λ_4 and given q^α the p^α are independent Gaussians with some means and isotropic variance determined by λ_i, $i = 1,2,3,4$. The variance is denoted by T, the means of the Gaussians by $u = \{u_i, 1 \le i \le 3\}$ and the density of the point process by ρ. (ρ, U, T) is another set of descriptions for our point processes and the mapping $\lambda \to (\rho, U, T)$ is one to one, invertible and smooth on the good set. The total energy per volume

$$e = \lim_{Q \uparrow R^3} \left\{ \frac{1}{2} \sum_\alpha \sum_i \chi_Q(q^\alpha)(p_i^\alpha)^2 + \frac{1}{2} \sum_\alpha \sum_\beta \chi_Q(q^\alpha)\chi_Q(q^\beta) V(q^\alpha - q^\beta) \right\}$$

can be computed as

$$e = \rho \left(\frac{1}{2} \sum_{i=1}^{3} u_i^2 + \frac{3}{2} T \right) + f(\rho, T)$$

so that (ρ, U, e) is another possible set of parameters. The Euler equations we have written are for this set of parameters.

One of the main questions to be answered is the relationship if any between the Hamiltonian equations and the Euler equations. We would like to make a connection along the following lines. If we solve the Euler equations with smooth solutions that remain in the good set then we have smooth functions for the parametric functions $\lambda_i(t,x)$, $i = 0,1,2,3,4$ as well. We can then define a probability measure $q_\epsilon(t, q^\alpha, p^\alpha)$ on $(T^3 \times R^3)^N$ by

$$\frac{1}{Z_\epsilon} \exp \Big[\sum \lambda_0(t, q^\alpha) + \sum \lambda_i(t, q^\alpha) p_i^\alpha \\ - \frac{1}{2} \sum_\alpha \lambda_4(t, q^\alpha) \{ \sum_i |p_i^\alpha|^2 + \sum_\beta V(\frac{q^\alpha - q^\beta}{\epsilon}) \Big]$$

where Z_ϵ is again the normalization constant.

We can write the Liouville operator

$$L_\epsilon = - \sum_{i,\alpha} p_i^\alpha \frac{\partial}{q_i^\alpha} + \sum_{i,\alpha,\beta} \frac{1}{\epsilon} V_i(\frac{q^\alpha - q^\beta}{\epsilon}) \frac{\partial}{\partial p_i^\alpha}$$

and solve the equation

$$\frac{\partial f_\epsilon^t}{\partial t} = L_\epsilon f_\epsilon^t$$
$$\text{with } f_\epsilon^t|_{t=0} = g(0, q^\alpha, p^\alpha) .$$

If we can show that f_ϵ^t is close to g_ϵ^t in the interval $[0,T]$ where the solution to the Euler equations are defined this would be a good theorem. We have to make two modifications before we can prove such a result. First of all we have difficulty with high velocities and are unable to control it by truncation. The Euler equation has a cubic term in the velocity and our entropy method forces us to exponentiate it and the cubic exponent is bad news even for the Gaussian. We therefore modify the Hamiltonian from $\frac{1}{2}\|p\|^2$ to a function $\phi(p)$ which is smooth, convex and has a bounded gradient. The velocities are then uniformly bounded. This of course changes the thermodynamics and the Euler equation. But the concepts are parallel. The more serious modification is the need for noise.

We need a minimal amount of noise and this is added by modifying the Liouville operator with the addition of a second order term to make it Fokker-Planck or parabolic. It is done carefully as follows. Suppose p^α and p^β are the momenta of two particles, then in any motion four quantities are conserved. The total momenta and energy for the two particles. This leaves a family two dimensional surface for (p^α, p^β) to slide on and still observe the conservation laws. We take a finite collection of smooth vector fields X_j, $j = 1, 2, \ldots, \ell$ and define the operator

$$L_{\alpha,\beta} = \sum_j X_j^* X_j$$

We take enough so that $L_{\alpha,\beta}$ is elliptic on the 2-dimensional manifolds. The modification of the Liouville operator is the following

$$\mathcal{A}_\epsilon = L_\epsilon + \theta(\epsilon) \sum_{\alpha,\beta} \psi\left(\frac{q^\alpha - q^\beta}{\epsilon}\right) L_{\alpha,\beta}$$

This means there is a noisy exchange of momenta and energy between pairs of particles which are within an order "ϵ" of each other. $\psi(q)$ is a rapidly decaying function. $\theta(\epsilon) \to \infty$ as $\epsilon \to 0$ by $\epsilon\theta(\epsilon) \to 0$. It is therefore weak relative to the Liouville term which is of order ϵ. Remember that we have speeded up time. In original time the strength $\theta(\epsilon)$ will be such that $\theta(\epsilon) \ll 1$, but $\theta(\epsilon) \gg \epsilon$. If we now solve the degenerate heat equation

$$\frac{\partial f_\epsilon^t}{\partial t} = \mathcal{A}_\epsilon f_\epsilon^t$$
$$\text{with } f_\epsilon^0 = g_\epsilon^0$$

then the result is

$$\lim_{\epsilon \to 0} \sup_{0 \le t \le T} \epsilon^3 \int f_\epsilon^t \log \frac{f_\epsilon^t}{g_\epsilon^t} \, dp^\alpha \, dq^\alpha = 0 \, .$$

The proof is carried out in a manner analogous to the earlier case. The Dirichlet form coming from the second order operator is sufficient to tell us that locally we have for velocities mixtures of Gaussian distributions. We can then use the Hamiltonian Dynamics to prove a local ergodic theorem. Once we have the local ergodic theorem the relative entropy method will work, for we have evaded the troublesome issue of velocity truncation.

6 Some additional models and remarks.

We shall examine some more models to see how the method we have developed applies. The first one is the interacting Brownian motion on the circle S. We take N copies of S and define a diffusion on the N-torus by

$$dx_i = -N \sum_j V'(N(x_i - x_j)) + d\beta_i(t)$$

The infinitesimal generator is

$$\mathcal{L}_N = \frac{1}{2}\sum \frac{\partial^2}{\partial x_i^2} - N\sum_j V'(N(x_i - x_j))\frac{\partial}{\partial x_i}$$

Here $V(x)$ is a repulsive short range potential on R and $V(N(x-y))$ is the rescaled pair potential on S that defines the interaction. We assume that $V(x) = V(-x)$. The diffusion has a unique invariant measure given by the density

$$\frac{1}{Z_N}\exp\left[-\frac{1}{2}\sum_{i,j} V(N(x_i - x_j))\right]$$

L_N is reversible with respect to the above weight. If we start with an initial density $g(x_1), \ldots, g(x_N)$ on $S^{(N)}$ and solve the Fokker-Planck equation

$$\frac{\partial f_N^t}{\partial t} = \mathcal{L}_N^* f_N^t \,, \quad f_N^t|_{t=0} = g(x_1)\cdots g(x_N)$$

then for $t > 0$ we get the solution $f_N^t(x_1, \ldots, x_N)$ which is a symmetric function. The one particle density

$$g_N(t, x_1) = \int f_N^t(x_1, \ldots, x_N)\, dx_2 \cdots dx_N$$

does not satisfy any simple equation for finite N. However, as $N \to \infty$, $g_N(t,x)$ has a limit $g(t,x)$ that satisfies a nonlinear equation

$$\frac{\partial g}{\partial t} = \frac{1}{2}(P(g))_{xx} \quad \text{for} \quad t > 0\,,\ x \in S$$

with the initial condition

$$g(t,x)|_{t=0} = g(x)\,.$$

The function $P(\rho)$ of $\rho \geq 0$ is determined by thermodynamic considerations of the one dimensional system with pair interaction $V(x-y)$.

The problem can be handled by both methods that we have outlined if the dimension is 1. However if $d > 1$, the relative entropy method works if we stay away from phase transitions. In principle the phase transitions should yield only zero diffusion and except for the possibility of the nonlinear diffusion coefficient thus having some zeros nothing else should change. This has been carried out in the lattice Ginzburg-Landau case, but not in the interacting Brownian motion case. The relative entropy method runs into trouble when there are phase transitions. The first method using Dirichlet forms runs into trouble if $d > 1$ in the interacting Brownian motion case.

Another example is to modify our Ginzburg-Landau model so that the Dirichlet form is

$$\frac{N^2}{2}\int \sum a(x_i, x_{i+1})\left(\frac{\partial u}{\partial x_i} - \frac{\partial u}{\partial x_{i+1}}\right)^2 e^{-\sum \phi(x_i)}\, dx_1 \ldots dx_N$$

The operator takes the form

$$\frac{N^2}{2}\sum a(x_i, x_{i+1})\left(\frac{\partial}{\partial x_i} - \frac{\partial}{\partial x_{i+1}}\right)^2 - \frac{N^2}{2}\sum W(x_i, x_{i+1})\left(\frac{\partial}{\partial x_i} - \frac{\partial}{\partial x_{i+1}}\right)$$

Because $W(x_i, x_{i+1})$ is not a difference of the form $\phi'(x_i) - \phi'(x_{i+1})$ like in the case when $a \equiv 1$ we can sum by parts only once. Instead of looking at terms of the form

$$\frac{1}{N}\sum J''(\frac{i}{N})\phi'(x_i)$$

we are forced to stare at

$$\sum J'(\frac{i}{N})\, W(x_i, x_{i+1})$$

One has to work hard to convince ourselves why

$$\int_0^t \sum J'(\frac{i}{N})\, W(x_i(s), x_{i+1}(s)\,)\, ds$$

is of order 1 and work even order to identify it in terms of averages of x_i. The final equation takes the form

$$\frac{\partial m}{\partial t} = \frac{1}{2}(\hat{a}(m)(h'(m)\,)_x\,)_x \quad \text{on} \quad [0, \infty) \times S$$

The diffusion coefficient $\hat{a}(m)$ is computed as follows. We consider the product measure P_m on R^∞ given by

$$dP_m = \prod \frac{1}{M(\lambda)} e^{\lambda x_i - \phi(x_i)}\, dx_i$$

where $\lambda = h'(m)$. Consider an arbitrary smooth function $F(x_{i-k}, x_{i+k})$ depending on $2k+1$ variables and "look at" the sum

$$\sum_{-\infty < i < \infty} F(x_{i-k}, \ldots, x_{i+k}) = U_F$$

U_F is of course totally ridiculous but

$$\xi_F = \frac{\partial U_F}{\partial x_1} - \frac{\partial U_F}{\partial x_2}$$

is well defined in $L_2(P_m)$ and has $E^{P_m}[\xi_F] = 0$. The span of all such ξ_F in $L_2(P_M)$ as k and F vary is denoted by $\mathcal{H}_0$. Then $1 \notin H_0$ and

$$\hat{a}(m) = \inf_{\xi \in \mathcal{H}_0} E^{P_m}\, a(x_1, x_2)\,(1 - \xi)^2$$

is the diffusion coefficient. If $a \equiv 1$, then $\hat{a}(m) = 1$ because $\xi = 0$ cannot be improved.

Remarks and References.

The place where the entropy methods are introduced by Guo-Papanicolaou and Varadhan is [5]. This is carried out for the Ginzburg-Landau model on a periodic lattice. This model has been studied by Fritz by different methods in [3]. Fritz has also modified these methods in [4] to the infinite volume case. The use of the entropy method to study large deviations and weak perturbations has been carried out by Donsker and Varadhan in [2] for the Ginzburg-Landau model and by Kipnis-Olla-Varadhan in [6] for the symmetric random walk with simple exclusion. The relative entropy method has been used by Yau in [14] for the Ginzburg-Landau model and by Olla-Varadhan-Yau in [7] for the modified Hamiltonian system. The Ginzburg-Landau model with phase transitions has been studied by Rezakhanlou in [9]. The interacting Brownian motion in one dimension was studied in [12] by Varadhan. In [10] Spohn had studied earlier fluctuations in equilibrium for the same model. The socalled nongradient version of the Ginzburg-Landau model was studied in [13] by Varadhan. Similar ideas have been used by Quastel [8] to study particles of two distinct colors in the simple exclusion model.

We have limited ourselves to entropy methods that work well in the diffusive scaling limit. The asymmetric random walk which leads to a convective limit has been studied in great detail by other methods, especially coupling methods. For an exhaustive list of references and a survey, two excellent sources are the recent texts by DeMasi and Presutti [1] and Spohn [11].

References

[1] DeMasi, A. and Presutti, E., *Mathematical Methods for Hydrodynamical Limits*, Lecture Notes in Mathematics, No. 1501, Springer-Verlag, Berlin - Heidelberg - New York, 1991.

[2] Donsker, M. D. and Varadhan, S. R. S., *Large deviations from a hydrodynamic scaling limit*, Comm. Pure Appl. Math. 42 (1989) 243–270.

[3] Fritz, J., *On the hydrodynamic limit of a scalar Ginzburg-Landau lattice model*, IMA Vol. 9, Berlin - Heidelberg - New York, Springer-Verlag, 1987.

[4] Fritz, J., *On the diffusive nature of entropy flow in infinite systems; remarks to a paper by Guo-Papanicolaou-Varadhan*, Comm. Math. Phys. 133 (1990) 331–352.

[5] Guo, M. Z., Papanicolaou, G. C. and Varadhan, S. R. S., *Nonlinear diffusion limit for a system with nearest neighbor interaction*, Comm. Math. Phys. 118 (1988) 31–59.

[6] Kipnis, C., Olla, S. and Varadhan, S. R. S., *Hydrodynamics and large deviation for simple exclusion process*, Comm. Pure Appl. Math. 42 (1989) 115–137.

[7] Olla, S., Varadhan, S. R. S. and Yau, H. T., *Hydrodynamics for Hamiltonian systems with small noise*, in preparation.

[8] Quastel, J., *Diffusion of color in the simple exclusion process*, Comm. Pure Appl. Math. 45 (1992) 623–680.

[9] Rezakhanlou, F., *Hydrodynamic limit for a system with finite range interactions*, Comm. Math. Phys. 129 (1990) 448–480.

[10] Spohn, H., *Equilibrium fluctuations for interacting Brownian particles*, Comm. Math. Phys. 103 (1986) 1–33.

[11] Spohn, H., *Large Scale Dynamics of Interacting Particles*, Texts and Monographs in Physics, Springer-Verlag, Berlin - Heidelberg - New York, 1991.

[12] Varadhan, S. R. S., *Scaling limits for interacting diffusions*, Commun. Math. Phys. 135 (1991) 313–353.

[13] Varadhan, S. R. S., *Nonlinear diffusion limit for a system with nearest neighbor interactions II*, Proc. Taniguchi Symp., 1990, Kyoto.

[14] Yau, H. T., *Relative entropy and the hydrodynamics of Ginzburg-Landau models*, Lett. Math. Phys. 22 (1991) 63–80.

C.I.M.E. Session on "Nonequilibrium Problems in Many-Particle Systems"

List of Participants

V. ALLEVA, Via P. Maroncelli 45, 00149 Roma

L. ARLOTTI, Dip. di Mat., Campus Universitario, Trav. 200 Re David 4, 70125 Bari

A. ARNOLD, Dept. of Math., Purdue Univ., West Lafayette, IN 47907

J. BATT, Math. Inst., Univ. München, Theresienstrasse 39, D-8 München 2

D. BENEDETTO, Dip. di Mat., Univ. di Roma Tor Vergata, Viale della Ricerca Scientifica, 00133 Roma

M.S. BERNABEI, Strada delle Grazie 18, 62020 Monte San Martino (MC)

L. BERTINI-MALGARINI, Dip. di Mat., Università di Roma Tor Vergata, Viale della Ricerca Scientifica, 00133 Roma

L. BONAVENTURA, Courant Institute, 251 Mercer Street, New York, NY 10012

F. BOUCHUT, CNRS/PMMS, 3A Av. de la Recherche Scientifique, 45071 Orléans Cedex 2

P. BROMAN, Dept. of Math., Chalmers Univ. of Tech., S-41296 Göteborg

G. BUSONI, Dip. di Mat., Viale Morgagni 67/A, 50134 Firenze

E. CAGLIOTI, Dip. di Mat., Univ. La Sapienza, P.le Aldo Moro 2, 00185 Roma

S. CAPRINO, Via Valdagno 31, 00191 Roma

G.L. CARAFFINI, Dip. di Mat., Via D'Azeglio 85, 43100 Parma

C.-C. CHANG, Dept. of Math., National Taiwan Univ., Taipei, Taiwan 107, R.O.C.

Qi CHEN, Fachb. Math., Univ. Kaiserslautern, P.O.Box 3049, D-6750 Kaiserslautern

E. DAMBROGIO, Dip. di Mat., P.le Europa 1, 34124 Trieste

J. DOLBEAULT, Lab. de Phys. Quant., Univ. Paul Sabatier, 118 route de Narbonne 31062 Toulouse Cedex

R. ESPOSITO, Dip. di Mat., Univ. di Roma Tor Vergata, Viale della Ricerca Scientifica, 00133 Roma

G. FROSALI, Dip. di Mat., Via Brecce Bianche, 60131 Ancona

F. FUCHS, Lab. de Math., Univ. de Nice, Parc Valrose, 06034 Nice

E. GABETTA, Dip. di Mat., Univ. di Milano, Via C. Saldini 50, 20133 Milano

N. IANIRO, Dip. di Metodi e Modelli Matematici, VIa S. Scarpa 10, 00161 Roma

U. KRAUSE, Inst. fur Ang. Math. und Stoch., Mohrenstrasse 39, D-1086 Berlin

M.I. LOFFREDO, Dip. di Mat., Univ. di Siena, Via del Capitano 15, 53100 Siena

R. MARRA, P.le Montesquieu 28/G, 00137 Roma

N. MAUSER, RB Mathematik, MA 6-2, TU-Berlin, Str. d. 17. Juni 136, D-1000 Berlin 12

B. MONTAGNINI, Dip. di Costruzioni Mecc. e Nucleari, Via Diotisalvi 2, 56100 Pisa

W. MOROKOFF, IMA, 514 Vincent Hall, Univ. of Minnesota, Minneapolis, MN 55455

O. MUSCATO, Dip. di Mat., Città Universitaria, Viale A. Doria 6, 95125 Catania

A. NOURI, Lab. de Math., Univ. de Nice, Parc Valrose, F-06034 Nice

B. PERTHAME, Dept. de Math., Univ. d'Orléans, BP 6759, 45067 Orléans Cedex 2

A. PULVIRENTI, Dip. di Mat., Univ. di Ferrara, Via Machiavelli 35, 44100 Ferrara

G. REIN, Math. Inst. der Univ. München, Theresienstr. 39, 8 München 2

B. RUBINO, Scuola Normale Superiore, Piazza dei Cavalieri 7, 56100 Pisa

B. RUDIGER, Dip. di Mat., Univ. di Roma Tor Vergata, Viale della Ricerca Scientifica, 00133 Roma

W. SACK, FB Mathematik, Univ. Kaiserslautern, P.O. Box 3049, D-6750 Kaiserslautern

C. SGARRA, Dip. di Mat., Politecnico di Milano, P.za L. da Vinci 32, 20133 Milano

G. TOSCANI, Dip. di Mat., Univ. di Ferrara, Via Machiavelli 35, 44100 Ferrara

L. TRIOLO, Dip. di Mat., Univ. di Roma Tor Vergata, Viale della Ricerca Scientifica, 00133 Roma

J. VERDINA, Dept. of Math., California State Univ., Long Beach, Cal. 90840

W. WAGNER, Inst. fur Ang. Anal. und Stoch., Hausvogteiplatz 5-7, D-O-1086 Berlin

H. WATANABE, Dept. of Appl. Math., Okayama Univ. of Science, Ridai-cho 11, Okayama 70(

B. WENNBERG, Dept. of Math., Chalmers Univ. of Tech., S-41296 Göteborg

M. WIDDER, Inst. for Theor. Phys., Altenbergerstr. 69, A-4045 Linz

A. YAMNAHAKKI, Lab. de Math., Univ. de Nice, Parc Valrose, F-06034 Nice

FONDAZIONE C.I.M.E.
CENTRO INTERNAZIONALE MATEMATICO ESTIVO
INTERNATIONAL MATHEMATICAL SUMMER CENTER

"Dirichlet Forms"

is the subject of the First 1992 C.I.M.E. Session.

The Session, sponsored by the Consiglio Nazionale delle Ricerche and the Ministero dell'Università e della Ricerca Scientifica e Tecnologica, will take place under the scientific direction of Prof. GIANFAUSTO DELL'ANTONIO (Università di Roma, La Sapienza), and Prof. UMBERTO MOSCO (Università di Roma, La Sapienza), at Villa Monastero, Varenna (Lake of Como), **from June 8 to June 19, 1992.**

Courses

a) **Parabolic Harnack inequalities and the behavior of fundamental solutions.** (5 lectures in English)
Prof. Eugen FABES (University of Minnesota)

Outline

A. Gaussian estimates for the fundamental solution of nondegenerate parabolic operators - applications to Harnack inequalities. (Ref. 4,6,7).

B. The Harnack inequality implies estimates for the fundamental solution - the nondegenerate and degenerate cases. (Ref. 1,2,5).

C. Gaussian estimates for heat flows on Riemannian manifolds. (Ref. 3).

References

1. Aronson, D.G., Bounds for the fundamental solution of a parabolic equation, Bulletin of the AMS 73 (1967), 890-896.
2. Chiarenza, F.M., Serapioni, R.P., A Harnack inequality for degenerate parabolic equations, Comm. in PDE 9 (1984), 719-749.
3. Davies, E.B., Heat Kernels and Spectral Theory, Cambridge Tracts in Mathematics 92, Cambridge University Press, 1990.
4. Fabes, E.B., Stroock, D.W., A new proof of Moser's parabolic Harnack inequality via the old ideas of Nash, Arch. Rat. Mech. Anal. 96 (1986), 327-338.
5. Gutierrez, C.E., Wheeden, R.L., Bounds for the fundamental solution of degenerate parabolic equations, to appear.
6. Moser, J., A Harnack inequality for parabolic differential equations, Comm. Pure and Applied Math. 17 (1964), 101-134, and also Correction to "A Harnack inequality for parabolic differential equations", Comm. Pure and Applied Math. 20 (1967), 232-236.
7. Nash, J., Continuity of solutions of parabolic and elliptic equations, Amer. J. Math. 80 (1958), 931-954.

b) **General Theory of Dirichlet forms: Part II.** (6 lectures in English).
Prof. Masatoshi FUKUSHIMA (Osaka University)

A short outline of the course

Chapter 1. Dirichlet forms in finite dimensional analysis.
In this chapter, are given some specific but basic examples of Dirichlet forms related to the finite dimensional analysis, e.g., spatially homogeneous Dirichlet forms, Dirichlet forms on fractal sets and Dirichlet forms generated by plurisubharmonic functions. As applications, quasi-everywhere convergences of Fourier series, point recurrence properties and spectral dimensions for fractal sets and characterizations of pluripolar sets are discussed.

Chapter 2. Stochastic analysis by additive functionals.
In this chapter, the theory of additive functionals associated with the regular Dirichlet form is presented; positive

continuous additive functionals and smooth measures, decompositions of additive functionals and their relations to Dirichlet forms. Various applications with specific emphasis on the roles of martingale additive functionals are given.

References

1. M. Fukushima, Dirichlet forms and Markov processes, North-Holland and Kodan-sha, 1980.
2. M. Fukushima and M. Takeda, A transformation of a symmetric Markov process and the Donsker-Varadhan theory, Osaka J. Math. 21 (1984), 311-326.
3. M. Fukushima and M. Okada, On Dirichlet forms for plurisubharmonic functions, Acta Math. 54 (1987), 171-213.
4. M. Fukushima, Dirichlet forms, diffusion processes and spectral dimensions for nested fractals, in "Ideas and Methods in Mathematical Analysis, Stochastics, and Applications, In Memory of R. Hoegh-Krohn, Vol. 1", Albeverio, Penstad, Holden, Lindstrom (eds.), Cambridge Univ. Press, to appear.
5. M. Takeda, On a martingale method for symmetric diffusion processes and its applications, Osaka J. Math. 26 (1989), 606-623.

c) **Logarithmic Sobolev inequalities over finite and infinite dimensional spaces.** (6 lectures in English).
Prof. Leonard GROSS (Cornell University).

Outline

I. Logarithmic Sobolev inequalities in L^2.
  A. The standard Gaussian L.S. inequality on R^n
  B. L.S. generators in L^2
  C. General properties
    1. Semiboundedness of perturbations
    2. Additivity
    3. Mass gap

II. Hypercontractive, supercontractive and ultracontractive semigroups.
  A. Definitions
  B. L.S. generators of index p in $(0,\infty)$
  C. Equivalence of L.S. generators with strong contraction properties
  D. Examples

III. Logarithmic Sobolev inequalities for Dirichlet forms.
  A. Index 2 implies index p
  B. Example (Nelson's best estimates for the number operator)

IV. Survey of applications.
  A. Heat kernel bounds - Davies et al.
  B. Semiboundedness of Hamiltonians

V. Some recent applications to Schrödinger operators loop groups.

Prerequisites

1. Spectral theorem for unbounded self-adjoint operator
2. The Beurling-Deny theorem - as in M. Fukushima, "Dirichlet Forms and Markov Processes", North-Holland, New York, 1980.

d) **Potential theory of non-divergence form elliptic operators.** (5 lectures in English)
Prof. Carlos KENIG (University of Chicago)

Outline

The purpose of these lectures will be to describe the potential theory of non divergence form elliptic equations and their adjoints, highlighting the connection with a natural degenerate Dirichlet form. We will study both local behavior and boundary behavior, including a Wiener test.

References

[B1] Bauman, P., Positive solutions of elliptic equations in nondivergence form and their adjoints, Arkiv für Mat. 22 (1984), 153-173.
[B2] Bauman, P., A Wiener test for nondivergence structure, second order elliptic equations, Indiana U. Math. J. 4 (1985), 825-844.
[F,G,M,S] Fabes, E., Garofalo, N., Marin Malave, S. and Salsa, S., Fatou theorems for some nonlinear elliptic equations, Revista Mat. Iberoamericana, Vol. 4 (1988), 227-251.
[F,S] Fabes, E. and Stroock, D., The L^p-integrability of Green's functions and fundamental solutions for elliptic and parabolic equations, Duke Math. J. 51 (1984), 977-1016.
[P] Pucci, C., Limitazioni per soluzioni di equazioni ellittiche, Ann. Mat. Pura ed Appl. 74 (1966), 15-30.
[S] Safonov, M.V., Harnack's inequality for elliptic equations and the Hölder property of their solutions, J. Soviet Math. 21 (1983), 851-863.

e) **General theory of Dirichlet forms: Part I.** (6 lectures in English).
Prof. Michael RÖCKNER (Universität Bonn)

Outline

The purpose of this part of the course is both to give an introduction to the theory of (not necessarily symmetric) Dirichlet forms and to describe their significance in infinite dimensional analysis. The first half will consist of a presentation of the basic theory on general state spaces and includes of the following topics:

1. Underlying L^2-theory and contraction properties
2. Potential theory of Dirichlet forms
3. Necessary and sufficient conditions for the existence of an associated Markov process
4. Compactification

The second half of the lectures is devoted to applications with a special emphasis on examples with infinite dimensional state spaces. It will cover the following topics:

5. Closability of classical (pre-) Dirichlet forms on topological vector spaces
6. Tightness of capacities
7. Existence and uniqueness of solutions for stochastic differential equations in infinite dimensional space
8. Girsanov transform in infinite dimensions

References

1. Albeverio, S., Röckner, M.: Stochastic differential equations in infinite dimensions: solutions via Dirichlet forms. Probab. Th. Rel. Fields 89 (1991), 347-386.
2. Fukushima, M.: Dirichlet forms and Markov processes. Amsterdam-Oxford-New York, North Holand (1980).
3. Ma, Z., Röckner, M.: An introduction to the theory of (non-symmetric) Dirichlet forms. Monograph, to appear.
4. Silverstein, M.L.: Symmetric Markov Processes. Lecture Notes in Math. 426. Berlin-Heidelberg-New York, Springer (1974).
5. Röckner, M., Zhang, T.S.: On uniqueness of generalized Schrödinger operators and applications. To appear in J. Funct. Anal.

f) **Dirichlet forms and ergodic properties.** (6 lectures in English).
Prof. Daniel W. STROOCK (MIT)

Outline

The course will emphasize the application of Dirichlet forms to the long time analysis of various stochastic processes. The topics which will be covered are as follows.

1. General comments about spectral gaps, Sobolev, and logarithmic Sobolev inequalities and their relationship to ergodic phenomena.
2. Analysis of the "simulated annealing" procedure.
3. Criteria for the existence of a logarithmic Sobolev inequality.
4. Preliminary discussion of Gibbs states and Glauber dynamics.
5. Conditions under which a Gibbs state admits a logarithmic Sobolev inequality.

References

Chapter VI of Large deviations by J.-D. Deuschel and D. Stroock, Academic Press, (1989).
R. Holley, D. Stroock, J. of Stat. Physics, 46 (1987), 1159-1194.
E. A. Carlen, S. Kusuoka, D.W. Stroock, Ann. Inst. H. Poincaré, Probab. et Stat., Supl. au n° 2, 1987, 245-287.
R. Holley, D. Stroock, Commun. Math. Phys., 115 (1988), 553-569.
R. Holley, S. Kusuoka, D. Stroock, J. Funct. Anal. 83 (1989), 333-334.
J. D. Deuschel, D.W. Stroock, J. Funct. Anal. 92 (1990), 30-48.
P. Diaconis, D. Stroock, Ann. of Appl. Prob. 1 (1991), 36-61.
D.W. Stroock, B. Zegarlinski, The logarithmic Sobolev inequality for continuous spin systems on a lattice, to appear J. Funct. Anal.
D.W. Stroock, B. Zegarlinski, The equivalence of the logarithmic Sobolev inequality and the Dobrushin-Shlosman mixing condition, to appear Comm. Math. Phys.

FONDAZIONE C.I.M.E
CENTRO INTERNAZIONALE MATEMATICO ESTIVO
INTERNATIONAL MATHEMATICAL SUMMER CENTER

"D-Modules and Representation Theory"

is the subject of the Second 1992 C.I.M.E. Session.

The Session, sponsored by the Consiglio Nazionale delle Ricerche and by the Ministero dell'Università e della Ricerca Scientifica e Tecnologica, will take place under the scientific direction of Prof. GIUSEPPE ZAMPIERI (Università di Padova) and Prof. ANDREA D'AGNOLO (Università di Padova) at Università di Venezia, Ca' Foscari, Venezia from **June 12 to June 20, 1992.**

Courses

a) **Formule de l'indice relative.** (5 lectures in English)
Prof. Louis BOUTET DE MONVEL (Université de Paris VI)

Résumé

Soient X et Y deux variétés analytiques, $f: Y \to X$ une application analytique, M un D_y-module cohérent ("bien filtré"). Le théoréme de l'indice décrit l'élément de K-théorie $[f_* M]$ associé à l'image directe $f_* M$ en fonction de [M] lorsque M est relativement elliptique (condition qui assure que $[f_* M]$ est cohérent et bien filtré, d'aprés Houzel et Schapira).

Les conférences comprendront deux parties:

1) une de rappels des éléments de K-théorie nécéssaires à la description de la formule, en particulier de la K-théorie à supports et de la description pour celle ci du théoréme de périodicité au moyen d'opérateurs de Toeplitz.
2) une description de la partie pertinente de la théorie des D-modules et des images directes de D-modules permettant d'énoncer le théoréme d'indice et d'en décrire la démonstration.

Reférences bibliographiques

[A] M.F. Atiyah, K-theory, Benjamin, Amsterdam.
[Bo] A. Borel et al., Algebraic D-modules, Perspect. in Math. n. 2, Academic Press.
[B-M] L. Boutet de Monvel, B. Malgrange, Le théorème de l'indice relatif. Ann. Sc. E.N.S., à paraître.
[BM1] L. Boutet de Monvel, On the index of Toeplitz operators of several complex variables, Inventiones Math. 50 (1979), 249-272.
[BM2] L. Boutet de Monvel, The index of almost elliptic systems. E. De Giorgi Colloquium, Research notes in Math. 125, Pitman 1985, 17-29.
[GRo] A. Grothendieck, SGA V, théorie des intersections et théorème de Riemann-Roch, Lecture Notes in Math. 225, Springer Verlag (1971)
[Hi] F. Hirzebruch, Neue topologische Methoden in der algebraische Geometrie. Springer Verlag, Berlin.
[Hö] L. Hörmander, The Analysis of Linear Partial Differential Operators, Vol. III et IV, Grundlehren der Math. Wiss. 124.
[H-Sch] Ch. Houzel, P. Schapira, Images directes de modules différentiels, C.R.A.S. 298 (1984), 461-464.
[M] B. Malgrange, Sur les images directes de D-modules, Manuscripta Math. 50 (1985), 49-71.
[K] M. Kashiwara, Cours Université Paris Nord, Birkhauser 1983.
[K-K-S] M. Kashiwara, T. Kawai, M. Sato, Microfunctions and pseudo-differential equations, Lecture Notes 287 (1973), Springer Verlag.

b) **Quantized Enveloping Algebras and Their Representations.** (10 lectures in English).
Prof. Corrado DE CONCINI (SNS, Pisa) and Prof. Claudio PROCESI (Univ. Roma La Sapienza).

Outline

1. Poisson Lie Groups.
2. Quantized enveloping algebras. The universal R-matrix. The center of quantized enveloping algebras, the Harish-Chandra isomorphism.
3. Representation for generic q. Their complete reducibility and their classification and characters.
4. Specializations at roots of unity. The center at roots of 1. The subalgebra Z_0 of the center as the coordinate ring of a Poisson algebraic group.
5. Representations at roots of 1. Their relations with coadjoint orbits.
6. The divided power algebra of Lusztig at a root of 1. Its representations.

c) **Index theorems for constructible sheaves and D-modules.** (5 lectures in English).
Prof. Pierre SCHAPIRA (Université Paris Nord)

Outline

1) Subanalytic stratifications and constructible sheaves
2) Lagrangian cycles
3) Characteristic cycle of constructible sheaves and index theorem
4) Applications to holonomic D-modules
5) Elliptic pairs and open problems

The main reference

M. Kashiwara and P. Schapira: Sheaves on manifolds. Grundlehren der Math. Wiss., Springer Verlag, 292 (1990).

One may also consult:

M. Kashiwara: Systems of microdifferential equations. Progress in Math. 34, Birkhäuser (1983).
P. Schapira: Microdifferential systems in the complex domain. Grundlehren der Math. Wiss., Springer Verlag, 269 (1985).

d) **Cohomologie équivariante et théorèmes d'indice.** (5 lectures in English).
Prof. Michèle VERGNE (ENS, Paris)

Outline

- Cohomologie de De Rham d'une variété
- Fibres, superconnections, classes caractéristiques
- Cohomologie équivariante d'une variété. Fibrés équivariants, classes caractéristiques en cohomologie équivariante
- Cohomologie équivariante et indice de l'opérateur de Dirac
- Cohomologie équivariante et théorème de l'indice pour les opérateurs transversalement elliptiques
- Représentations des groupes compacts. Formule d'Hermann Weyl et formule de Kirillov pour les caractères.

Références bibliographiques:

Berline-Getsler-Vergne, Heat kernels and Dirac operators, à paraitre Springer.

LIST OF C.I.M.E. SEMINARS

		Publisher
1954 -	1. Analisi funzionale	C.I.M.E.
	2. Quadratura delle superficie e questioni connesse	"
	3. Equazioni differenziali non lineari	"
1955 -	4. Teorema di Riemann-Roch e questioni connesse	"
	5. Teoria dei numeri	"
	6. Topologia	"
	7. Teorie non linearizzate in elasticità, idrodinamica,aerodinamica	"
	8. Geometria proiettivo-differenziale	"
1956 -	9. Equazioni alle derivate parziali a caratteristiche reali	"
	10. Propagazione delle onde elettromagnetiche	"
	11. Teoria della funzioni di più variabili complesse e delle funzioni automorfe	"
1957 -	12. Geometria aritmetica e algebrica (2 vol.)	"
	13. Integrali singolari e questioni connesse	"
	14. Teoria della turbolenza (2 vol.)	"
1958 -	15. Vedute e problemi attuali in relatività generale	"
	16. Problemi di geometria differenziale in grande	"
	17. Il principio di minimo e le sue applicazioni alle equazioni funzionali	"
1959 -	18. Induzione e statistica	"
	19. Teoria algebrica dei meccanismi automatici (2 vol.)	"
	20. Gruppi, anelli di Lie e teoria della coomologia	"
1960 -	21. Sistemi dinamici e teoremi ergodici	"
	22. Forme differenziali e loro integrali	"
1961 -	23. Geometria del calcolo delle variazioni (2 vol.)	"
	24. Teoria delle distribuzioni	"
	25. Onde superficiali	"
1962 -	26. Topologia differenziale	"
	27. Autovalori e autosoluzioni	"
	28. Magnetofluidodinamica	"

1963 - 29. Equazioni differenziali astratte "
30. Funzioni e varietà complesse "
31. Proprietà di media e teoremi di confronto in Fisica Matematica "

1964 - 32. Relatività generale "
33. Dinamica dei gas rarefatti "
34. Alcune questioni di analisi numerica "
35. Equazioni differenziali non lineari "

1965 - 36. Non-linear continuum theories "
37. Some aspects of ring theory "
38. Mathematical optimization in economics "

1966 - 39. Calculus of variations Ed. Cremonese, Firenze
40. Economia matematica "
41. Classi caratteristiche e questioni connesse "
42. Some aspects of diffusion theory "

1967 - 43. Modern questions of celestial mechanics "
44. Numerical analysis of partial differential equations "
45. Geometry of homogeneous bounded domains "

1968 - 46. Controllability and observability "
47. Pseudo-differential operators "
48. Aspects of mathematical logic "

1969 - 49. Potential theory "
50. Non-linear continuum theories in mechanics and physics and their applications "
51. Questions of algebraic varieties "

1970 - 52. Relativistic fluid dynamics "
53. Theory of group representations and Fourier analysis "
54. Functional equations and inequalities "
55. Problems in non-linear analysis "

1971 - 56. Stereodynamics "
57. Constructive aspects of functional analysis (2 vol.) "
58. Categories and commutative algebra "

1972 -	59. Non-linear mechanics		"
	60. Finite geometric structures and their applications		"
	61. Geometric measure theory and minimal surfaces		"
1973 -	62. Complex analysis		"
	63. New variational techniques in mathematical physics		"
	64. Spectral analysis		"
1974 -	65. Stability problems		"
	66. Singularities of analytic spaces		"
	67. Eigenvalues of non linear problems		"
1975 -	68. Theoretical computer sciences		"
	69. Model theory and applications		"
	70. Differential operators and manifolds		"
1976 -	71. Statistical Mechanics		Ed Liguori, Napoli
	72. Hyperbolicity		"
	73. Differential topology		"
1977 -	74. Materials with memory		"
	75. Pseudodifferential operators with applications		"
	76. Algebraic surfaces		"
1978 -	77. Stochastic differential equations		"
	78. Dynamical systems		Ed Liguori, Napoli and Birhäuser Verlag
1979 -	79. Recursion theory and computational complexity		"
	80. Mathematics of biology		"
1980 -	81. Wave propagation		"
	82. Harmonic analysis and group representations		"
	83. Matroid theory and its applications		"
1981 -	84. Kinetic Theories and the Boltzmann Equation	(LNM 1048)	Springer-Verlag
	85. Algebraic Threefolds	(LNM 947)	"
	86. Nonlinear Filtering and Stochastic Control	(LNM 972)	"
1982 -	87. Invariant Theory	(LNM 996)	"
	88. Thermodynamics and Constitutive Equations	(LN Physics 228)	"
	89. Fluid Dynamics	(LNM 1047)	"

1983 -	90. Complete Intersections	(LNM 1092)	"
	91. Bifurcation Theory and Applications	(LNM 1057)	"
	92. Numerical Methods in Fluid Dynamics	(LNM 1127)	"
1984 -	93. Harmonic Mappings and Minimal Immersions	(LNM 1161)	"
	94. Schrödinger Operators	(LNM 1159)	"
	95. Buildings and the Geometry of Diagrams	(LNM 1181)	"
1985 -	96. Probability and Analysis	(LNM 1206)	"
	97. Some Problems in Nonlinear Diffusion	(LNM 1224)	"
	98. Theory of Moduli	(LNM 1337)	"
1986 -	99. Inverse Problems	(LNM 1225)	"
	100. Mathematical Economics	(LNM 1330)	"
	101. Combinatorial Optimization	(LNM 1403)	"
1987 -	102. Relativistic Fluid Dynamics	(LNM 1385)	"
	103. Topics in Calculus of Variations	(LNM 1365)	"
1988 -	104. Logic and Computer Science	(LNM 1429)	"
	105. Global Geometry and Mathematical Physics	(LNM 1451)	"
1989 -	106. Methods of nonconvex analysis	(LNM 1446)	"
	107. Microlocal Analysis and Applications	(LNM 1495)	"
1990 -	108. Geoemtric Topology: Recent Developments	(LNM 1504)	"
	109. H Control Theory	(LNM 1496)	"
	110. Mathematical Modelling of Industrical Processes	(LNM 1521)	"
1991 -	111. Topological Methods for Ordinary Differential Equations	(LNM 1537)	"
	112. Arithmetic Algebraic Geometry	to appear	"
	113. Transition to Chaos in Classical and Quantum Mechanics	to appear	"
1992 -	114. Dirichlet Forms	to appear	"
	115. D-Modules and Representation Theory	to appear	"
	116. Nonequilibrium Problems in Many-Particle Systems	(LNM 1551)	"

Vol. 1460: G. Toscani, V. Boffi, S. Rionero (Eds.), Mathematical Aspects of Fluid Plasma Dynamics. Proceedings, 1988. V, 221 pages. 1991.

Vol. 1461: R. Gorenflo, S. Vessella, Abel Integral Equations. VII, 215 pages. 1991.

Vol. 1462: D. Mond, J. Montaldi (Eds.), Singularity Theory and its Applications. Warwick 1989, Part I. VIII, 405 pages. 1991.

Vol. 1463: R. Roberts, I. Stewart (Eds.), Singularity Theory and its Applications. Warwick 1989, Part II. VIII, 322 pages. 1991.

Vol. 1464: D. L. Burkholder, E. Pardoux, A. Sznitman, Ecole d'Eté de Probabilités de Saint- Flour XIX-1989. Editor: P. L. Hennequin. VI, 256 pages. 1991.

Vol. 1465: G. David, Wavelets and Singular Integrals on Curves and Surfaces. X, 107 pages. 1991.

Vol. 1466: W. Banaszczyk, Additive Subgroups of Topological Vector Spaces. VII, 178 pages. 1991.

Vol. 1467: W. M. Schmidt, Diophantine Approximations and Diophantine Equations. VIII, 217 pages. 1991.

Vol. 1468: J. Noguchi, T. Ohsawa (Eds.), Prospects in Complex Geometry. Proceedings, 1989. VII, 421 pages. 1991.

Vol. 1469: J. Lindenstrauss, V. D. Milman (Eds.), Geometric Aspects of Functional Analysis. Seminar 1989-90. XI, 191 pages. 1991.

Vol. 1470: E. Odell, H. Rosenthal (Eds.), Functional Analysis. Proceedings, 1987-89. VII, 199 pages. 1991.

Vol. 1471: A. A. Panchishkin, Non-Archimedean L-Functions of Siegel and Hilbert Modular Forms. VII, 157 pages. 1991.

Vol. 1472: T. T. Nielsen, Bose Algebras: The Complex and Real Wave Representations. V, 132 pages. 1991.

Vol. 1473: Y. Hino, S. Murakami, T. Naito, Functional Differential Equations with Infinite Delay. X, 317 pages. 1991.

Vol. 1474: S. Jackowski, B. Oliver, K. Pawałowski (Eds.), Algebraic Topology, Poznań 1989. Proceedings. VIII, 397 pages. 1991.

Vol. 1475: S. Busenberg, M. Martelli (Eds.), Delay Differential Equations and Dynamical Systems. Proceedings, 1990. VIII, 249 pages. 1991.

Vol. 1476: M. Bekkali, Topics in Set Theory. VII, 120 pages. 1991.

Vol. 1477: R. Jajte, Strong Limit Theorems in Noncommutative L_2-Spaces. X, 113 pages. 1991.

Vol. 1478: M.-P. Malliavin (Ed.), Topics in Invariant Theory. Seminar 1989-1990. VI, 272 pages. 1991.

Vol. 1479: S. Bloch, I. Dolgachev, W. Fulton (Eds.), Algebraic Geometry. Proceedings, 1989. VII, 300 pages. 1991.

Vol. 1480: F. Dumortier, R. Roussarie, J. Sotomayor, H. Żoładek, Bifurcations of Planar Vector Fields: Nilpotent Singularities and Abelian Integrals. VIII, 226 pages. 1991.

Vol. 1481: D. Ferus, U. Pinkall, U. Simon, B. Wegner (Eds.), Global Differential Geometry and Global Analysis. Proceedings, 1991. VIII, 283 pages. 1991.

Vol. 1482: J. Chabrowski, The Dirichlet Problem with L^2-Boundary Data for Elliptic Linear Equations. VI, 173 pages. 1991.

Vol. 1483: E. Reithmeier, Periodic Solutions of Nonlinear Dynamical Systems. VI, 171 pages. 1991.

Vol. 1484: H. Delfs, Homology of Locally Semialgebraic Spaces. IX, 136 pages. 1991.

Vol. 1485: J. Azéma, P. A. Meyer, M. Yor (Eds.), Séminaire de Probabilités XXV. VIII, 440 pages. 1991.

Vol. 1486: L. Arnold, H. Crauel, J.-P. Eckmann (Eds.), Lyapunov Exponents. Proceedings, 1990. VIII, 365 pages. 1991.

Vol. 1487: E. Freitag, Singular Modular Forms and Theta Relations. VI, 172 pages. 1991.

Vol. 1488: A. Carboni, M. C. Pedicchio, G. Rosolini (Eds.), Category Theory. Proceedings, 1990. VII, 494 pages. 1991.

Vol. 1489: A. Mielke, Hamiltonian and Lagrangian Flows on Center Manifolds. X, 140 pages. 1991.

Vol. 1490: K. Metsch, Linear Spaces with Few Lines. XIII, 196 pages. 1991.

Vol. 1491: E. Lluis-Puebla, J.-L. Loday, H. Gillet, C. Soulé, V. Snaith, Higher Algebraic K-Theory: an overview. IX, 164 pages. 1992.

Vol. 1492: K. R. Wicks, Fractals and Hyperspaces. VIII, 168 pages. 1991.

Vol. 1493: E. Benoît (Ed.), Dynamic Bifurcations. Proceedings, Luminy 1990. VII, 219 pages. 1991.

Vol. 1494: M.-T. Cheng, X.-W. Zhou, D.-G. Deng (Eds.), Harmonic Analysis. Proceedings, 1988. IX, 226 pages. 1991.

Vol. 1495: J. M. Bony, G. Grubb, L. Hörmander, H. Komatsu, J. Sjöstrand, Microlocal Analysis and Applications. Montecatini Terme, 1989. Editors: L. Cattabriga, L. Rodino. VII, 349 pages. 1991.

Vol. 1496: C. Foias, B. Francis, J. W. Helton, H. Kwakernaak, J. B. Pearson, H_∞-Control Theory. Como, 1990. Editors: E. Mosca, L. Pandolfi. VII, 336 pages. 1991.

Vol. 1497: G. T. Herman, A. K. Louis, F. Natterer (Eds.), Mathematical Methods in Tomography. Proceedings 1990. X, 268 pages. 1991.

Vol. 1498: R. Lang, Spectral Theory of Random Schrödinger Operators. X, 125 pages. 1991.

Vol. 1499: K. Taira, Boundary Value Problems and Markov Processes. IX, 132 pages. 1991.

Vol. 1500: J.-P. Serre, Lie Algebras and Lie Groups. VII, 168 pages. 1992.

Vol. 1501: A. De Masi, E. Presutti, Mathematical Methods for Hydrodynamic Limits. IX, 196 pages. 1991.

Vol. 1502: C. Simpson, Asymptotic Behavior of Monodromy. V, 139 pages. 1991.

Vol. 1503: S. Shokranian, The Selberg-Arthur Trace Formula (Lectures by J. Arthur). VII, 97 pages. 1991.

Vol. 1504: J. Cheeger, M. Gromov, C. Okonek, P. Pansu, Geometric Topology: Recent Developments. Editors: P. de Bartolomeis, F. Tricerri. VII, 197 pages. 1991.

Vol. 1505: K. Kajitani, T. Nishitani, The Hyperbolic Cauchy Problem. VII, 168 pages. 1991.

Vol. 1506: A. Buium, Differential Algebraic Groups of Finite Dimension. XV, 145 pages. 1992.

Vol. 1507: K. Hulek, T. Peternell, M. Schneider, F.-O. Schreyer (Eds.), Complex Algebraic Varieties. Proceedings, 1990. VII, 179 pages. 1992.

Vol. 1508: M. Vuorinen (Ed.), Quasiconformal Space Mappings. A Collection of Surveys 1960-1990. IX, 148 pages. 1992.

Vol. 1509: J. Aguadé, M. Castellet, F. R. Cohen (Eds.), Algebraic Topology - Homotopy and Group Cohomology. Proceedings, 1990. X, 330 pages. 1992.

Vol. 1510: P. P. Kulish (Ed.), Quantum Groups. Proceedings, 1990. XII, 398 pages. 1992.

Vol. 1511: B. S. Yadav, D. Singh (Eds.), Functional Analysis and Operator Theory. Proceedings, 1990. VIII, 223 pages. 1992.

Vol. 1512: L. M. Adleman, M.-D. A. Huang, Primality Testing and Abelian Varieties Over Finite Fields. VII, 142 pages. 1992.

Vol. 1513: L. S. Block, W. A. Coppel, Dynamics in One Dimension. VIII, 249 pages. 1992.

Vol. 1514: U. Krengel, K. Richter, V. Warstat (Eds.), Ergodic Theory and Related Topics III, Proceedings, 1990. VIII, 236 pages. 1992.

Vol. 1515: E. Ballico, F. Catanese, C. Ciliberto (Eds.), Classification of Irregular Varieties. Proceedings, 1990. VII, 149 pages. 1992.

Vol. 1516: R. A. Lorentz, Multivariate Birkhoff Interpolation. IX, 192 pages. 1992.

Vol. 1517: K. Keimel, W. Roth, Ordered Cones and Approximation. VI, 134 pages. 1992.

Vol. 1518: H. Stichtenoth, M. A. Tsfasman (Eds.), Coding Theory and Algebraic Geometry. Proceedings, 1991. VIII, 223 pages. 1992.

Vol. 1519: M. W. Short, The Primitive Soluble Permutation Groups of Degree less than 256. IX, 145 pages. 1992.

Vol. 1520: Yu. G. Borisovich, Yu. E. Gliklikh (Eds.), Global Analysis – Studies and Applications V. VII, 284 pages. 1992.

Vol. 1521: S. Busenberg, B. Forte, H. K. Kuiken, Mathematical Modelling of Industrial Process. Bari, 1990. Editors: V. Capasso, A. Fasano. VII, 162 pages. 1992.

Vol. 1522: J.-M. Delort, F. B. I. Transformation. VII, 101 pages. 1992.

Vol. 1523: W. Xue, Rings with Morita Duality. X, 168 pages. 1992.

Vol. 1524: M. Coste, L. Mahé, M.-F. Roy (Eds.), Real Algebraic Geometry. Proceedings, 1991. VIII, 418 pages. 1992.

Vol. 1525: C. Casacuberta, M. Castellet (Eds.), Mathematical Research Today and Tomorrow. VII, 112 pages. 1992.

Vol. 1526: J. Azéma, P. A. Meyer, M. Yor (Eds.), Séminaire de Probabilités XXVI. X, 633 pages. 1992.

Vol. 1527: M. I. Freidlin, J.-F. Le Gall, Ecole d'Eté de Probabilités de Saint-Flour XX – 1990. Editor: P. L. Hennequin. VIII, 244 pages. 1992.

Vol. 1528: G. Isac, Complementarity Problems. VI, 297 pages. 1992.

Vol. 1529: J. van Neerven, The Adjoint of a Semigroup of Linear Operators. X, 195 pages. 1992.

Vol. 1530: J. G. Heywood, K. Masuda, R. Rautmann, S. A. Solonnikov (Eds.), The Navier-Stokes Equations II – Theory and Numerical Methods. IX, 322 pages. 1992.

Vol. 1531: M. Stoer, Design of Survivable Networks. IV, 206 pages. 1992.

Vol. 1532: J. F. Colombeau, Multiplication of Distributions. X, 184 pages. 1992.

Vol. 1533: P. Jipsen, H. Rose, Varieties of Lattices. X, 162 pages. 1992.

Vol. 1534: C. Greither, Cyclic Galois Extensions of Commutative Rings. X, 145 pages. 1992.

Vol. 1535: A. B. Evans, Orthomorphism Graphs of Groups. VIII, 114 pages. 1992.

Vol. 1536: M. K. Kwong, A. Zettl, Norm Inequalities for Derivatives and Differences. VII, 150 pages. 1992.

Vol. 1537: P. Fitzpatrick, M. Martelli, J. Mawhin, R. Nussbaum, Topological Methods for Ordinary Differential Equations. Montecatini Terme, 1991. Editors: M. Furi, P. Zecca. VII, 218 pages. 1993.

Vol. 1538: P.-A. Meyer, Quantum Probability for Probabilists. X, 287 pages. 1993.

Vol. 1539: M. Coornaert, A. Papadopoulos, Symbolic Dynamics and Hyperbolic Groups. VIII, 138 pages. 1993.

Vol. 1540: H. Komatsu (Ed.), Functional Analysis and Related Topics, 1991. Proceedings. XXI, 413 pages. 1993.

Vol. 1541: D. A. Dawson, B. Maisonneuve, J. Spencer, Ecole d' Eté de Probabilités de Saint-Flour XXI - 1991. Editor: P. L. Hennequin. VIII, 356 pages. 1993.

Vol. 1542: J.Fröhlich, Th.Kerler, Quantum Groups, Quantum Categories and Quantum Field Theory. VII, 431 pages. 1993.

Vol. 1543: A. L. Dontchev, T. Zolezzi, Well-Posed Optimization Problems. XII, 421 pages. 1993.

Vol. 1544: M.Schürmann, White Noise on Bialgebras. VII, 146 pages. 1993.

Vol. 1545: J. Morgan, K. O'Grady, Differential Topology of Complex Surfaces. VIII, 224 pages. 1993.

Vol. 1546: V. V. Kalashnikov, V. M. Zolotarev (Eds.), Stability Problems for Stochastic Models. Proceedings, 1991. VIII, 229 pages. 1993.

Vol. 1547: P. Harmand, D. Werner, W. Werner, M-ideals in Banach Spaces and Banach Algebras. VIII, 387 pages. 1993.

Vol. 1548: T. Urabe, Dynkin Graphs and Quadrilateral Singularities. VI, 233 pages. 1993.

Vol. 1549: G. Vainikko, Multidimensional Weakly Singular Integral Equations. XI, 159 pages. 1993.

Vol. 1551: L. Arkeryd, P. L. Lions, P.A. Markowich, S.R. S. Varadhan. Nonequilibrium Problems in Many-Particle Systems. Montecatini, 1992. Editors: C. Cercignani, M. Pulvirenti. VII, 158 pages 1993.

Vol. 1552: J. Hilgert, K.-H. Neeb, Lie Semigroups and their Applications. XII, 315 pages. 1993.

Vol. 1553: J.-L- Colliot-Thélène, J. Kato, P. Vojta. Arithmetic Algebraic Geometry. Editor: E. Ballico. VII, 223 pages. 1993.

Vol. 1554: A. K. Lenstra, H. W. Lenstra, Jr. (Eds.), The Development of the Number Field Sieve. VIII, 131 pages. 1993.

Vol. 1555: O. Liess, Conical Refraction and Higher Microlocalization. X, 389 pages. 1993.